Environmental Engineering
Practice PE Exams

Third Edition

R. Wane Schneiter, PhD, PE, DEE

Professional Publications, Inc. • Belmont, CA

How to Locate Errata and Other Updates for This Book

At Professional Publications, we do our best to bring you error-free books. But when errors do occur, we want to make sure that you know about them so they cause as little confusion as possible.

A current list of known errata and other updates for this book is available on the PPI website at **www.ppi2pass.com**. From the website home page, click on "Errata." We update the errata page as often as necessary, so check in regularly. You will also find instructions for submitting suspected errata. We are grateful to every reader who takes the time to help us improve the quality of our books by pointing out an error.

ENVIRONMENTAL ENGINEERING PRACTICE PE EXAMS
Third Edition

Printing History

edition number	printing number	update
2	1	New edition.
2	2	Minor corrections.
3	1	New edition.

Copyright © 2004 by Professional Publications, Inc. All rights reserved. No part of this publication may be reproduced, stored in a retrieval system, or transmitted, in any form or by any means, electronic, mechanical, photocopying, recording, or otherwise, without the prior written permission of the publisher.

Printed in the United States of America

Professional Publications, Inc.
1250 Fifth Avenue, Belmont, CA 94002
(650) 593-9119
www.ppi2pass.com

Current printing of this edition: 1

Library of Congress Cataloging-in-Publication Data
Schneiter, R. W.
    Environmental engineering practice PE exams / R. Wane Schneiter.--3rd ed.
       p. cm.
    Includes bibliographical references.
    ISBN 1-59126-001-9 (pbk.)
    1. Environmental engineering--Examinations, questions, etc. 2. National Council of Examiners for Engineering and Surveying--Examinations--Study guides. I. Title.

TD157.C36 2004
628--dc22
                                                               2003062224

Table of Contents

PREFACE v

INTRODUCTION vii

SAMPLE EXAMINATION 1
    Morning Session 1
    Afternoon Session 11
    Solutions 21

SAMPLE EXAMINATION 2
    Morning Session 39
    Afternoon Session 49
    Solutions 59

SAMPLE EXAMINATION 3
    Morning Session 77
    Afternoon Session 87
    Solutions 97

RESOURCES 117

Preface

The National Council of Examiners for Engineering and Surveying (NCEES) prepares the Principles and Practice of Engineering (PE) examination for environmental engineering from problems submitted by professional engineers representing consulting, government, and industry. The exam provides a uniform tool for local licensing boards to assess the competency of engineers practicing within their jurisdictions. NCEES does not provide copies of past exams, and it keeps exam problems confidential so that actual problems included on any NCEES exam cannot be included in *Environmental Engineering Practice PE Exams* or any other study guide. However, NCEES does identify the general subject areas covered on the exam.

Within the limitations imposed by NCEES, *Environmental Engineering Practice PE Exams* presents a broad range of problems relevant to those that may be encountered in the practice of environmental engineering. Organized as three separate practice exams of 100 problems each, *Environmental Engineering Practice PE Exams* addresses those general subject areas identified by NCEES and covers a broad range of topics, both conceptual and practical, with varying levels of difficulty, and in a variety of forms. Although you will not encounter any problems on the exam exactly like those presented in this book, the problems presented here are believed to be representative of the type and difficulty of those you will encounter on the exam.

Environmental Engineering Practice PE Exams should be used to practice solving engineering problems in a test format and to evaluate your overall preparedness for taking the exam. You should not rely on *Environmental Engineering Practice PE Exams* as your only study guide to prepare for the exam. You should instead combine its use with the use of other study references. Review environmental engineering principles and concepts before attempting the sample problems, and then use these sample problems to determine your weaknesses and to direct subsequent study. This will enable you to focus your efforts where they will be most productive, to practice solving exam-like problems, and to compile and organize materials you may wish to use while taking the exam.

Solutions presented for each practice problem may represent only one of several alternative methods for obtaining a correct answer. It may also be possible that an alternative method of solving a problem will produce a different, but still appropriate, answer.

Although great care was exercised in preparing *Environmental Engineering Practice PE Exams* to ensure that representative problems were included and that they were solved correctly, some errors may exist. I ask for your assistance in improving subsequent editions by bringing any errors to my attention and by providing comments regarding the applicability of specific problems to those you encounter on the exam.

Good luck!

R. Wane Schneiter, PhD, PE, DEE
Lexington, VA

Introduction

EXAM ORGANIZATION

The PE exam consists of two parts, a morning and an afternoon session, each lasting four hours. In completing the exam you will need to provide answers to 100 multiple-choice problems, 50 in the morning and 50 in the afternoon. Most problems are completely independent, but some are grouped in sets of two to five problems, each set preceded by a problem statement. Four possible answers are provided for each problem, but only one of the possible answers is correct. One point is awarded for each correct answer. No partial credit is given, but no penalty is assessed for incorrect answers.

The morning and afternoon sessions will include problems that address topics in the following categories.

1. *Water.* Planning, research, development, project implementation, operations, and monitoring of waste, wastewater, storm water, and natural water systems.

2. *Solid and Hazardous Waste.* Planning, research, development, project implementation, operations, and monitoring of solid and hazardous waste systems.

3. *Air.* Planning, research, development, project implementation, operations, and monitoring of air systems. To include topics in pollution source, pollution control processes, and ambient air quality.

4. *Environmental Health, Safety, and Welfare.* Planning research, development, project implementation, operations, and monitoring of environmental health, safety, and welfare. To include topics in risk assessment, occupational and radiological health, fate and transport, and public health.

Also, some problems may require knowledge of engineering economics.

Both the morning and the afternoon sessions of the exam are open book. In general, any bound reference material is allowed, including (if bound) personal notes and sample calculations. Textbooks, handbooks, and other professional reference books are allowed. However, no writing tablets, scratch paper, or other unbound notes or materials are permitted. Mechanical pencils will be provided at the test site. Battery-operated, silent, nonprinting calculators are allowed. Calculators with communication or text editing abilities are not allowed. Each local jurisdiction defines specifically which materials you are permitted to bring with you to the exam, so check with them early to ensure that you are compiling acceptable materials for use during the exam.

The exam is intended to assess your personal competence without discussing or sharing information with others during the exam period. Do not expect to share any reference works or to communicate with others while taking the exam.

All solutions to the exam problems must be marked on the multiple-choice answer sheet provided. Any notes or calculations marked in the exam booklet will not be considered as part of a solution and will not be graded. The multiple-choice answer sheet is machine-graded and is the only record that will be scored.

EXAM SCORING

The entire eight-hour exam will require that you complete 100 multiple-choice problems. Each problem is worth one point, with no penalty for incorrect choices. However, if more than one choice is marked for a single problem, no points will be awarded. To pass the exam, you must score at least 70 points (70%).

EXAM PROCEDURE

Arrive at the exam location early enough to allow ample time for organizing yourself and any materials you plan to use during the exam. Prior to beginning the exam, a proctor will review exam procedures with you and distribute exam booklets. Do not open the booklet until instructed to do so. Answer sheets are enclosed with the exam booklet. Instructions, which you should read carefully, will be printed on the outside cover of the exam booklet. The proctor will review the procedure for marking your answers.

If you complete the exam with more than 30 minutes remaining, you may be allowed to leave the examination room. With less than 30 minutes remaining, you will likely be required to remain seated until the end of the exam so others are not disturbed. Regardless of when you complete the exam, you must return the numbered exam booklet with your answer sheets to the proctor prior to leaving.

For the latest information on exam changes, please check the PPI website at www.ppi2pass.com.

Instructions

Name: _____
     Last   First   Middle Initial

Do not enter solutions in the test booklet. Complete solutions must be entered on the answer sheet provided by your proctor.

This is an open-book examination. You may use textbooks, handbooks, and other bound references, along with a battery-operated, silent, nonprinting calculator. Unbound reference materials and notes, scratchpaper, and writing tablets are not permitted. You may not consult with or otherwise share any materials or information with others taking the exam.

You must work all 50 multiple-choice problems in the four-hour period allocated for the morning session. Each of the 50 problems is worth 1 point. No partial credit will be awarded. Your score will be based entirely on the responses marked on the answer sheet. You may use blank spaces in the exam booklet for scratch work. However, no credit will be awarded for work shown in margins or on other pages of the exam booklet. Mark only one answer to each problem.

Principles and Practice of Engineering Examination

MORNING SESSION
Sample Examination 1

1.	Ⓐ	Ⓑ	Ⓒ	Ⓓ	26.	Ⓐ	Ⓑ	Ⓒ	Ⓓ
2.	Ⓐ	Ⓑ	Ⓒ	Ⓓ	27.	Ⓐ	Ⓑ	Ⓒ	Ⓓ
3.	Ⓐ	Ⓑ	Ⓒ	Ⓓ	28.	Ⓐ	Ⓑ	Ⓒ	Ⓓ
4.	Ⓐ	Ⓑ	Ⓒ	Ⓓ	29.	Ⓐ	Ⓑ	Ⓒ	Ⓓ
5.	Ⓐ	Ⓑ	Ⓒ	Ⓓ	30.	Ⓐ	Ⓑ	Ⓒ	Ⓓ
6.	Ⓐ	Ⓑ	Ⓒ	Ⓓ	31.	Ⓐ	Ⓑ	Ⓒ	Ⓓ
7.	Ⓐ	Ⓑ	Ⓒ	Ⓓ	32.	Ⓐ	Ⓑ	Ⓒ	Ⓓ
8.	Ⓐ	Ⓑ	Ⓒ	Ⓓ	33.	Ⓐ	Ⓑ	Ⓒ	Ⓓ
9.	Ⓐ	Ⓑ	Ⓒ	Ⓓ	34.	Ⓐ	Ⓑ	Ⓒ	Ⓓ
10.	Ⓐ	Ⓑ	Ⓒ	Ⓓ	35.	Ⓐ	Ⓑ	Ⓒ	Ⓓ
11.	Ⓐ	Ⓑ	Ⓒ	Ⓓ	36.	Ⓐ	Ⓑ	Ⓒ	Ⓓ
12.	Ⓐ	Ⓑ	Ⓒ	Ⓓ	37.	Ⓐ	Ⓑ	Ⓒ	Ⓓ
13.	Ⓐ	Ⓑ	Ⓒ	Ⓓ	38.	Ⓐ	Ⓑ	Ⓒ	Ⓓ
14.	Ⓐ	Ⓑ	Ⓒ	Ⓓ	39.	Ⓐ	Ⓑ	Ⓒ	Ⓓ
15.	Ⓐ	Ⓑ	Ⓒ	Ⓓ	40.	Ⓐ	Ⓑ	Ⓒ	Ⓓ
16.	Ⓐ	Ⓑ	Ⓒ	Ⓓ	41.	Ⓐ	Ⓑ	Ⓒ	Ⓓ
17.	Ⓐ	Ⓑ	Ⓒ	Ⓓ	42.	Ⓐ	Ⓑ	Ⓒ	Ⓓ
18.	Ⓐ	Ⓑ	Ⓒ	Ⓓ	43.	Ⓐ	Ⓑ	Ⓒ	Ⓓ
19.	Ⓐ	Ⓑ	Ⓒ	Ⓓ	44.	Ⓐ	Ⓑ	Ⓒ	Ⓓ
20.	Ⓐ	Ⓑ	Ⓒ	Ⓓ	45.	Ⓐ	Ⓑ	Ⓒ	Ⓓ
21.	Ⓐ	Ⓑ	Ⓒ	Ⓓ	46.	Ⓐ	Ⓑ	Ⓒ	Ⓓ
22.	Ⓐ	Ⓑ	Ⓒ	Ⓓ	47.	Ⓐ	Ⓑ	Ⓒ	Ⓓ
23.	Ⓐ	Ⓑ	Ⓒ	Ⓓ	48.	Ⓐ	Ⓑ	Ⓒ	Ⓓ
24.	Ⓐ	Ⓑ	Ⓒ	Ⓓ	49.	Ⓐ	Ⓑ	Ⓒ	Ⓓ
25.	Ⓐ	Ⓑ	Ⓒ	Ⓓ	50.	Ⓐ	Ⓑ	Ⓒ	Ⓓ

Exam 1—Morning Session

SITUATION FOR PROBLEMS 1–4

An industrial wastewater effluent is treated using a complete mix-activated sludge process with both primary and secondary clarification. Two secondary clarifiers are used to produce an effluent with volatile suspended solids (VSS) of 60 mg/L. The total flow rate from the primary clarifiers is 60 000 m^3/d and contains 100 mg/L VSS. A single bioreactor is employed with a volume 8000 m^3. Biomass production in the bioreactor is 0.5 kg/m^3·d. The return volatile solids concentration is 12 000 mg/L. The specific gravity of the volatile solids is 1.0.

1. For a mixed liquor volatile suspended solids (MLVSS) of 1800 mg/L, what is the current total return solids flow rate?

(A) 500 m^3/d
(B) 4300 m^3/d
(C) 9600 m^3/d
(D) 12 000 m^3/d

2. For a return solids flow rate of 5000 m^3/d and a MLVSS of 2200 mg/L, what is the wasted solids flow rate from each of the secondary clarifiers?

(A) 1300 m^3/d per clarifier
(B) 3500 m^3/d per clarifier
(C) 4000 m^3/d per clarifier
(D) 5400 m^3/d per clarifier

3. For a wasted solids flow rate of 1000 m^3/d, what is the current total volume of wasted solids requiring disposal at 20% solids?

(A) 32 m^3/d
(B) 60 m^3/d
(C) 290 m^3/d
(D) 6400 m^3/d

4. What solids mass must be wasted daily to maintain the MLVSS at constant concentration?

(A) 1400 kg/d
(B) 1800 kg/d
(C) 4000 kg/d
(D) 6900 kg/d

SITUATION FOR PROBLEMS 5–7

A municipal sanitary sewer authority is experiencing infiltration and inflow (I/I) problems. The average biochemical oxygen demand (BOD) from municipal sources where the sewer enters the wastewater treatment plant is 135 mg/L. The BOD from industrial dischargers averages 348 mg/L at an average industrial flow rate of 8000 m^3/d. The total average annual flow from all sources to the wastewater treatment plant is 50 000 m^3/d.

Monitoring at manholes has produced the following I/I data for a section of the sewer system.

pipe no.	up manhole	down manhole	pipe diameter (mm)	pipe length (m)	average flow (L/d)
1	3	2	600	85	2500
2	2	7	600	46	2000
3	4	3	300	40	140
4	5	4	250	67	500
5	8	6	200	91	50
6	7	8	200	131	80
7	1	6	150	50	110

5. Which pipes require rehabilitation when judged against infiltration rates for new systems?

(A) 2
(B) 1 and 2
(C) 1, 2, and 4
(D) 1, 2, 4, and 6

6. The greatest benefit per cost unit would result from rehabilitating which lines?

(A) 2
(B) 1 and 2
(C) 1, 2, and 4
(D) 1, 2, 4, and 6

7. Assuming a typical municipal influent BOD$_5$ of 200 mg/L, what is the estimated flow from I/I to the entire sewer system?

(A) 14 000 m^3/d
(B) 22 000 m^3/d
(C) 30 000 m^3/d
(D) 34 000 m^3/d

SITUATION FOR PROBLEMS 8–10

The average annual wastewater flow for a combined commercial/residential development is 9,000,000 gal/yr.

8. What is the daily average wastewater flow based on the maximum month for the development assuming typical wastewater flow rates apply and conventional plumbing fixtures and appliances are used?

- (A) 31,000 gal/day
- (B) 35,000 gal/day
- (C) 45,000 gal/day
- (D) 68,000 gal/day

9. What is the incremental annual BOD mass loading rate to a wastewater treatment plant from the development based on average annual wastewater flow?

- (A) 3800 kg BOD/yr
- (B) 5000 kg BOD/yr
- (C) 7500 kg BOD/yr
- (D) 10 000 kg BOD/yr

10. What is the impact on the incremental annual BOD mass loading rate to a wastewater treatment plant if conventional plumbing fixtures and appliances are replaced with water conservation plumbing fixtures and appliances?

- (A) no change because the concentration will increase
- (B) increases because the concentration will increase
- (C) decreases because the volume will decrease
- (D) decreases because the volume and concentration will decrease

SITUATION FOR PROBLEMS 11–14

A wooded site in the eastern United States has been selected for residential development. Site area uses are as follows.

undeveloped	77 ha
rooftops, roads, driveways, etc.	33 ha
landscaping	82 ha

The storm distribution is type II, as represented by the following illustrations.

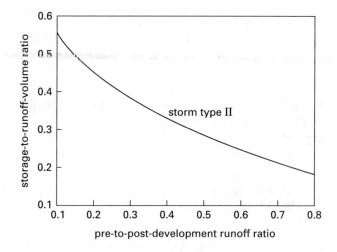

Illustration for Probs. 11–14

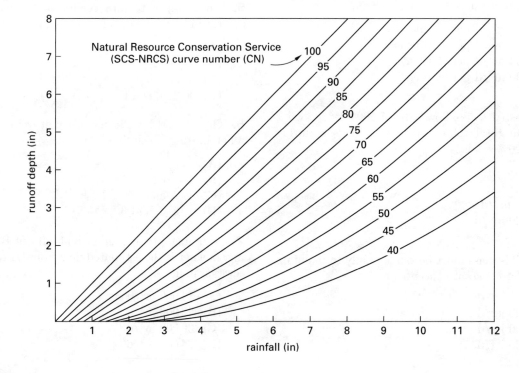

11. What is the required ratio of storage-to-runoff-volume when the ratio of pre-to-post-development runoff is 0.42?

(A) 0.29
(B) 0.32
(C) 0.38
(D) 0.48

12. What is the average runoff volume for the 25 yr storm?

(A) 0.060 cm
(B) 0.63 cm
(C) 1.3 cm
(D) 5.7 cm

13. What is the required storage volume for a runoff volume of 1.2 cm and a storage-to-runoff-volume ratio of 0.25?

(A) 0.25 cm
(B) 0.30 cm
(C) 1.2 cm
(D) 3.3 cm

14. What detention basin capacity is required for a storage volume of 0.95 cm?

(A) 4400 m^3
(B) 16 000 m^3
(C) 18 000 m^3
(D) 22 000 m^3

SITUATION FOR PROBLEMS 15–19

Type II settling column tests were performed for a flocculated suspension that, when completely mixed, contained a total suspended solids (TSS) concentration of 195 mg/L. The tests were conducted to provide design data for sedimentation basins required to treat 12 000 m^3/d of flow in four parallel units. The settling column tests provided the following results.

settling time (min)	TSS removal at indicated depth (%)					
	0.5 m	1.0 m	1.5 m	2.0 m	2.5 m	3.0 m
60	68	63	59	54	48	42
80	75	71	67	62	57	49
100	82	78	72	69	64	60
120	92	88	83	80	77	75

Select a settling zone depth of 2.5 m and a settling time of 100 min. The required sedimentation basin settling-zone length-to-width ratio is 3:1.

15. What is the overall TSS removal efficiency?

(A) 64%
(B) 73%
(C) 80%
(D) 85%

16. What is the overflow rate based on settling zones for rectangular basins?

(A) 0.38 m^3/m^2·h
(B) 0.75 m^3/m^2·h
(C) 0.96 m^3/m^2·h
(D) 1.5 m^3/m^2·h

17. What is the effect on efficiency if the settling zone depth is decreased to 1.5 m?

(A) decreases
(B) increases
(C) remains unchanged if the settling time remains unchanged
(D) remains unchanged if the overflow rate remains unchanged

18. What is the effect on efficiency if the settling time is increased to 120 min?

(A) decreases
(B) increases
(C) remains unchanged if the settling depth remains unchanged
(D) remains unchanged if the overflow rate remains unchanged

19. Assuming 80% efficiency, what is the daily volume of TSS removed if the solids content is 18%?

(A) 0.50 m^3/d
(B) 2.3 m^3/d
(C) 10 m^3/d
(D) 16 m^3/d

SITUATION FOR PROBLEMS 20–22

Dye tracer studies were conducted to define the flow characteristics of a reactor for a water treatment process. The reactor characteristics are defined by the following illustration.

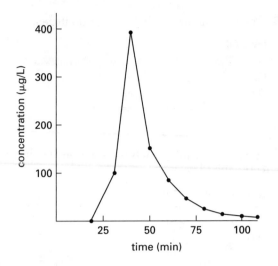

20. Does the reactor more closely model plug-flow or complete-mix conditions, neither, or a combination of the two?

(A) plug flow
(B) complete mix
(C) both plug flow and complete mix
(D) neither plug flow nor complete mix

21. For a slug tracer in an ideal plug-flow reactor, which of the following relationships is true?

(A) modal hydraulic residence time = theoretical hydraulic residence time
(B) modal hydraulic residence time = 0
(C) minimum hydraulic residence time = 0
(D) mean hydraulic residence time < theoretical hydraulic residence time

22. For a slug tracer in an ideal complete-mix reactor, which of the following relationships is true?

(A) modal hydraulic residence time = theoretical hydraulic residence time
(B) minimum hydraulic residence time = theoretical hydraulic residence time
(C) modal hydraulic residence time = 0
(D) mean hydraulic residence time > theoretical hydraulic residence time

SITUATION FOR PROBLEMS 23–26

An electronics component manufacturer generates wastewater that contains a volatile chemical. Air stripping was selected as the treatment method to remove the volatile chemical from the wastewater. The stripper design parameters are as follows.

wastewater flow rate	375 gal/min
chemical concentration in wastewater before treatment	2000 μg/L
chemical concentration in wastewater after treatment	20 μg/L
mass transfer coefficient	0.014/s
Henry's constant	0.18 (unitless)
stripping factor	3.5
hydraulic loading rate	30 gal/min ft^2
average water and air temperatures	25°C

23. What is the required air flow rate?

(A) 330 L/s
(B) 460 L/s
(C) 21 m^3/s
(D) 140 m^3/s

24. What is the number of transfer units (NTU) required?

(A) 1.6
(B) 3.7
(C) 6.0
(D) 7.8

25. What is the required height of each transfer unit (HTU)?

(A) 0.05 m
(B) 0.6 m
(C) 1.4 m
(D) 2.8 m

26. What is the required column diameter?

(A) 0.6 m
(B) 1.2 m
(C) 4.0 m
(D) 4.9 m

SITUATION FOR PROBLEMS 27–28

Design parameters for sedimentation basins at a small municipality are as follows.

flow rate	7500 m^3/d
overflow rate (type I)	1.8 m^3/m^2·h
settling zone depth	3.0 m
length to width ratio	3:1
weir overflow rate	14 m^3/m·h

27. What is the surface area of the sedimentation basin settling zone assuming two parallel tanks are used?

(A) 44 m²/tank
(B) 79 m²/tank
(C) 87 m²/tank
(D) 160 m²/tank

28. What is the weir length required for each of two sedimentation basins?

(A) 4.5 m/tank
(B) 11 m/tank
(C) 23 m/tank
(D) 140 m/tank

SITUATION FOR PROBLEMS 29–31

The following problems pertain to the three-pond system shown from a bird's eye view.

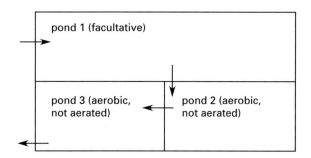

29. How could algae contributing to total suspended solids (TSS) discharges from pond 3 be controlled?

(A) constructing baffles in the pond
(B) excluding mechanical aeration from the pond
(C) locating the effluent structure to discharge from near the pond surface
(D) none of the above

30. If the influent to pond 1 contains 200 mg/L BOD, 15 mg/L total Kjeldahl nitrogen (TKN) and 4 mg/L total phosphorus (TP), what measures are justified for the pond system?

(A) augmentation of nitrogen
(B) augmentation of phosphorous
(C) provisions for removal of nitrogen and phosphorous
(D) no special measures for removal or augmentation are justified

31. How could the hydraulic efficiency of pond 2 be increased?

(A) installing surface aerators to provide complete mixing of pond contents
(B) changing the relative location of inlet and outlet structures
(C) installing baffles within the pond
(D) any combination of the above

SITUATION FOR PROBLEMS 32–35

A wastewater discharge to a river has the following characteristics.

flow rate	2.65 ft³/sec
temperature	88°F
dissolved oxygen	2.4 mg/L
BOD₅	198 mg/L at 20°C (k (base e) at 20°C = 0.26/d)

The river receiving the discharge has the following characteristics.

flow rate	68 ft³/sec
temperature	48°F
dissolved oxygen (DO) upstream of discharge point	8.9 mg/L
BOD₅ upstream of discharge point	2.1 mg/L
deoxygenation constant	0.16/d
reoxygenation constant	0.23/d

32. What is the BOD_u of the mixed flows?

(A) 9.4 mg/L
(B) 14 mg/L
(C) 100 mg/L
(D) 260 mg/L

33. For a mixed flow BOD_u at the discharge point of 10 mg/L, what is the DO concentration at a location 14 d downstream of the discharge?

(A) 1.6 mg/L
(B) 2.5 mg/L
(C) 9.1 mg/L
(D) 10 mg/L

34. What is the critical DO deficit corresponding to a critical time of 3.8 d for a mixed flow BOD_u at the discharge point of 10 mg/L?

(A) 4.0 mg/L
(B) 5.3 mg/L
(C) 6.6 mg/L
(D) 7.6 mg/L

35. What is the critical time for a mixed flow BOD_u at the discharge point of 10 mg/L?

(A) 1.7 d
(B) 2.6 d
(C) 3.4 d
(D) 6.4 d

SITUATION FOR PROBLEMS 36–38

A municipality operates an anaerobic digester to stabilize solids generated at its wastewater treatment plant. In an effort to recover some of the costs associated with operating the plant, the city is considering a proposal to collect the digester gases and recover the methane for sale as an energy source. The digester produces about 525,000 ft³ of gas daily. The gas contains 70% methane, 26% carbon dioxide, and 2% hydrogen sulfide, with the remaining 2% of gas consisting of water vapor and trace quantities of several other gases. Assume temperature and pressure of 20°C and 1 atm.

36. What is the daily mass of methane generated by the digester?

(A) 180 lbm/day
(B) 1000 lbm/day
(C) 16,000 lbm/day
(D) 22,000 lbm/day

37. What is the heating value of the methane per 1000 lbm of methane?

(A) 8.6×10^1 kJ/1000 ft³
(B) 9.3×10^2 kJ/1000 ft³
(C) 8.6×10^4 kJ/1000 ft³
(D) 1.0×10^6 kJ/1000 ft³

38. Assuming a heating value of 800 kJ/mol, what volume of methane gas at 20°C and 1 atm would be required if it was combusted to heat 25,000 gal of water from 10°C to 60°C?

(A) 340 ft³
(B) 4900 ft³
(C) 20,000 ft³
(D) 320,000 ft³

SITUATION FOR PROBLEMS 39–41

A metal plating process produces wastewater with the following characteristics.

flow	300 000 L/d
influent temperature	39°C
hydrochloric acid concentration	2.1 M

39. What is the reaction equation for neutralizing the hydrochloric acid with sodium hydroxide?

(A) $HOCl + NaOH \rightarrow H_2O + Na^+ + Cl^-$
(B) $HCl + NaOH \rightarrow H_2O + Na^+ + Cl^-$
(C) $H_2Cl + 2NaOH \rightarrow 2H_2O + 2Na^+ + Cl^-$
(D) $HCl_2 + NaOH \rightarrow H_2O + Na^+ + 2Cl^-$

40. Assuming a 1-to-1 molar ratio between the acid and the base, what is the daily mass of sodium hydroxide required to neutralize the acid?

(A) 630 kg/d
(B) 11 000 kg/d
(C) 14 000 kg/d
(D) 25 000 kg/d

41. What is the temperature of the wastewater after the addition of sodium hydroxide if the summed standard enthalpy for the reactants is −640 kJ/mol and for the products is −690 kJ/mol?

(A) 14°C
(B) 25°C
(C) 64°C
(D) 120°C

SITUATION FOR PROBLEMS 42–45

A community of 15 000 people is investigating municipal solid waste disposal options including landfilling and incineration. The 1.6 kg/person·d of solid waste generated by the community is characterized as follows.

waste component	mass (%)	moisture (%)	component discarded density (kg/m³)	component discarded energy (kJ/kg)	ash (%)
food	13	70	290	4650	5
glass	6	2	195	150	98
plastic	4	2	65	32600	10
paper	37	6	85	16750	6
cardboard	10	5	50	16300	5
textiles	1	10	65	17450	2.5
ferrous metal	8	3	320	700	98
nonferrous metal	2	2	160	700	96
wood	4	20	240	18600	1.5
yard clippings	15	60	105	6500	4.5

42. What is the daily discarded mass of solid waste requiring disposal?

(A) 6000 kg/d
(B) 12 000 kg/d
(C) 15 000 kg/d
(D) 24 000 kg/d

43. What is the discarded moisture content of the bulk waste?

(A) 18%
(B) 22%
(C) 36%
(D) 44%

44. What is the discarded density of the bulk waste?

(A) 79 kg/m^3
(B) 100 kg/m^3
(C) 120 kg/m^3
(D) 130 kg/m^3

45. What is the energy content of the discarded bulk waste without ash?

(A) 1.5×10^7 kJ/1000 kg
(B) 3.6×10^7 kJ/1000 kg
(C) 4.3×10^7 kJ/1000 kg
(D) 4.9×10^7 kJ/1000 kg

SITUATION FOR PROBLEMS 46–48

A municipal solid waste (MSW) is destined for a landfill. The municipality collects the waste from its 13 000 residents who each generate the waste at approximately 1.2 kg/person·d. The chemical composition of the waste, as discarded, is $C_{81}H_{179}O_{68}N$.

46. What is the methane production potential from decomposition of the discarded bulk waste?

(A) 730 kg CH_4/d
(B) 5000 kg CH_4/d
(C) 7700 kg CH_4/d
(D) 16 000 kg CH_4/d

47. What will be the primary constituents of the landfill gas?

(A) CH_4
(B) CH_4 and CO_2
(C) CH_4, CO_2, and NH_3
(D) CH_4, CO_2, NH_3, and H_2S

48. What is the landfill volume required for a 25 yr active life with a cover-to-waste ratio of 1:5 and a maximum compacted density of 1100 kg/m^3?

(A) 160×10^3 m^3
(B) 240×10^3 m^3
(C) 1300×10^3 m^3
(D) 2200×10^3 m^3

SITUATION FOR PROBLEMS 49–50

Incinerators are selected based on the type of materials requiring incineration and the air emissions limits, among other criteria.

49. What type of incinerator would be best suited for burning a mixed waste of dewatered biological sludge and shredded textiles?

(A) controlled air
(B) rotary kiln
(C) fluid bed
(D) multiple hearth

50. Which incinerator type presents the most difficulty in meeting air emission limits?

(A) controlled air
(B) rotary kiln
(C) fluid bed
(D) multiple hearth

Instructions

Name: _____
    Last   First   Middle Initial

Do not enter solutions in the test booklet. Complete solutions must be entered on the answer sheet provided by your proctor.

This is an open-book examination. You may use textbooks, handbooks, and other bound references, along with a battery-operated, silent, nonprinting calculator. Unbound reference materials and notes, scratchpaper, and writing tablets are not permitted. You may not consult with or otherwise share any materials or information with others taking the exam.

You must work all 50 multiple-choice problems in the four-hour period allocated for the afternoon session. Each of the 50 problems is worth 1 point. No partial credit will be awarded. Your score will be based entirely on the responses marked on the answer sheet. You may use blank spaces in the exam booklet for scratch work. However, no credit will be awarded for work shown in margins or on other pages of the exam booklet. Mark only one answer to each problem.

Principles and Practice of Engineering Examination

AFTERNOON SESSION
Sample Examination 1

51. (A) (B) (C) (D)   76. (A) (B) (C) (D)
52. (A) (B) (C) (D)   77. (A) (B) (C) (D)
53. (A) (B) (C) (D)   78. (A) (B) (C) (D)
54. (A) (B) (C) (D)   79. (A) (B) (C) (D)
55. (A) (B) (C) (D)   80. (A) (B) (C) (D)
56. (A) (B) (C) (D)   81. (A) (B) (C) (D)
57. (A) (B) (C) (D)   82. (A) (B) (C) (D)
58. (A) (B) (C) (D)   83. (A) (B) (C) (D)
59. (A) (B) (C) (D)   84. (A) (B) (C) (D)
60. (A) (B) (C) (D)   85. (A) (B) (C) (D)
61. (A) (B) (C) (D)   86. (A) (B) (C) (D)
62. (A) (B) (C) (D)   87. (A) (B) (C) (D)
63. (A) (B) (C) (D)   88. (A) (B) (C) (D)
64. (A) (B) (C) (D)   89. (A) (B) (C) (D)
65. (A) (B) (C) (D)   90. (A) (B) (C) (D)
66. (A) (B) (C) (D)   91. (A) (B) (C) (D)
67. (A) (B) (C) (D)   92. (A) (B) (C) (D)
68. (A) (B) (C) (D)   93. (A) (B) (C) (D)
69. (A) (B) (C) (D)   94. (A) (B) (C) (D)
70. (A) (B) (C) (D)   95. (A) (B) (C) (D)
71. (A) (B) (C) (D)   96. (A) (B) (C) (D)
72. (A) (B) (C) (D)   97. (A) (B) (C) (D)
73. (A) (B) (C) (D)   98. (A) (B) (C) (D)
74. (A) (B) (C) (D)   99. (A) (B) (C) (D)
75. (A) (B) (C) (D)   100. (A) (B) (C) (D)

Exam 1—Afternoon Session

SITUATION FOR PROBLEMS 51–52

An incinerator is to have a bed area loading of 2.3×10^6 kJ/m²·h for a waste stream of 500 kg/h with a heating value of 57 000 kJ/kg.

51. What is the minimum required incinerator bed area?

(A) 0.5 m²
(B) 4.3 m²
(C) 12 m²
(D) 26 m²

52. Which of the following are important factors for determining incinerator efficiency?

(A) burn temperature, oxygen level, turbulence, residence time
(B) burn temperature, viscosity, turbulence, residence time
(C) burn temperature, oxygen level, corrosivity, residence time
(D) burn temperature, oxygen level, flame velocity, turbulence

SITUATION FOR PROBLEMS 53–54

An electroplater produces wastewater at a continuous rate of 1.8 m³/min with Cr^{+6} (measured as CrO_3) at a concentration of 534 mg/L. Sodium bisulfite ($NaHSO_3$) and sulfuric acid (H_2SO_4) are selected to reduce the chromium from the hexavalent to the trivalent form. The chemical equation for the reduction reaction is

$$4CrO_3 + 6NaHSO_3 + 3H_2SO_4$$
$$\rightarrow 3Na_2SO_4 + 2Cr_2(SO_4)_3 + 6H_2O$$

Sodium bisulfite is available at 92% purity for $120/1000 kg, and sulfuric acid is available at 62% purity for $48/1000 kg.

53. What is the annual cost of sodium bisulfite required for reduction, assuming the facility operates 24 h/d, 365 d/yr?

(A) $69,000/yr
(B) $95,000/yr
(C) $100,000/yr
(D) $600,000/yr

54. What is the annual mass of reduced chromium that will require precipitation, assuming the facility operates 24 h/d, 365 d/yr?

(A) 1.3×10^5 kg/yr
(B) 5.1×10^5 kg/yr
(C) 9.9×10^5 kg/yr
(D) 2.0×10^6 kg/yr

SITUATION FOR PROBLEMS 55–56

An onshore facility uses an aboveground tank farm for bulk storage of fuel and lubricating oils. The tank farm, covered by a current spill prevention, control, and countermeasure (SPCC) plan, includes the following tanks.

tank no.	product stored	tank volume (gal)
1	no. 2 fuel oil	1,000,000
2	no. 2 fuel oil	1,000,000
3	no. 2 fuel oil	410,000
4	no. 2 fuel oil	650,000
5	SAE 60W lubricating oil	55,000
6	SAE 75W lubricating oil	18,000
7	SAE 90W lubricating oil	27,000

The bermed containment area surrounding the tanks must accommodate 110% of the volume of the largest tank plus the volume of rainfall produced by the 25 yr return, 24 hr rainfall depth of 14 in. The tank farm occupies 1.85 ac.

55. Which of the storage tanks are subject to inclusion in an SPCC plan?

(A) all of the tanks
(B) tanks 1 and 2 only
(C) tanks 1, 2, 3, and 4 only
(D) tanks 5, 6, and 7 only

56. If a berm were used for containment, what would be the required berm height assuming 1.5 ft freeboard?

(A) 2.9 ft
(B) 3.3 ft
(C) 4.5 ft
(D) 5.2 ft

SITUATION FOR PROBLEM 57

Air pollution sources include a broad range of activities, both man-made and natural, some of which are significant contributors to overall depleted air quality.

57. Which of the following is not a major source of air pollution in the United States?

(A) mobile sources such as cars, trucks, and buses
(B) stationary sources such as coal-fired power plants
(C) manufacturing associated with industries such as chemical and food processing
(D) open fires such as agricultural burning and forest/brush fires

SITUATION FOR PROBLEMS 58–59

The normal chemical composition of dry air is given in the following table.

component	concentration (%)
nitrogen	78.09
oxygen	20.94
argon	0.93
carbon dioxide	0.0315

The atmospheric pressure is 0.98 atm.

58. What is the partial pressure exerted by the argon?

(A) 0.0091 atm
(B) 0.098 atm
(C) 0.91 atm
(D) 1.1 atm

59. What is the apparent molecular weight of the air?

(A) 29 g/mol
(B) 36 g/mol
(C) 110 g/mol
(D) 140 g/mol

SITUATION FOR PROBLEM 60

Ground-level ozone was measured at 213 μg/m^3 in air at 1 atm and 24°C.

60. What is the ozone concentration in parts per million (ppm)?

(A) 0.087 ppm
(B) 0.099 ppm
(C) 0.11 ppm
(D) 0.14 ppm

SITUATION FOR PROBLEMS 61–62

Air quality monitoring includes measurements to assess compliance with the National Ambient Air Quality Standards (NAAQS).

61. What categories of measurements typically occur in air quality monitoring?

(A) emissions, ambient air, and meteorological
(B) emissions, meteorological, and human health
(C) emissions, ambient air, and environmental impact
(D) ambient air, human health, and environmental impact

62. What is the difference between primary and secondary standards, as used with the National Ambient Air Quality Standards (NAAQS), and primary and secondary air pollutants?

(A) Primary and secondary standards define the permissible levels of primary and secondary pollutants, respectively, that can exist in ambient air.
(B) Primary and secondary standards are regulatory levels to protect human health and prevent environmental damage, respectively. Primary and secondary pollutants relate to the emitted form of the pollutants.
(C) Primary and secondary standards refer to the level of treatment required by air pollution control equipment. Primary and secondary pollutants define the type of emissions from primary and secondary treatment equipment.
(D) There is no difference. They refer to the same thing.

SITUATION FOR PROBLEMS 63–65

An electrostatic precipitator (ESP) is being considered for use as an air pollution control (APC) process for a metal parts manufacturer. The air and particle characteristics are defined by the following parameters.

air flow rate requiring treatment	14 m³/s
average electric field	700 000 N/C
particulate concentration	20 g/m³
particle diameter	4.0 μm
temperature of gas stream	176°C

63. What is the particle drift velocity for the ESP?

(A) 0.23 m/s
(B) 0.32 m/s
(C) 1.4 m/s
(D) 1.9 m/s

64. For a drift velocity of 1.0 m/s, what is the plate area required for the ESP if an 80% efficiency is desired?

(A) 2 m²
(B) 10 m²
(C) 23 m²
(D) 70 m²

65. What is the daily mass of particulate removed by the ESP if operated at 80% efficiency?

(A) 3500 kg/d
(B) 7000 kg/d
(C) 14 000 kg/d
(D) 19 000 kg/d

SITUATION FOR PROBLEMS 66–67

Particulate matter with a mean diameter of 2.5 μm and a density of 0.58 g/cm³ is present in air at 1.3 atm and 30°C.

66. What is the particle-settling velocity?

(A) 0.0011 cm/s
(B) 0.010 cm/s
(C) 0.10 cm/s
(D) 0.42 cm/s

67. What is the Reynolds number for a particle-settling velocity of 0.10 m/s?

(A) 0.0020
(B) 0.015
(C) 0.020
(D) 0.084

SITUATION FOR PROBLEMS 68–70

Packed bed, vapor phase, granular activated carbon (GAC) adsorption is proposed for recovering a volatile organic solvent from a gas stream at 1 atm and 25°C. The solvent vapor is present at 180 ppm and the adsorption follows a Freundlich isotherm. The process parameters are presented as follows.

gas flow rate	4.2 m³/s
packed bed depth	60 cm
packed bed porosity	0.38
packed bed density	0.40 g/cm³
GAC mesh	6 × 12
regeneration cycle	1/d
isotherm intercept	0.126
isotherm slope	0.176
solvent molecular weight	78 g/mol

68. What is the GAC use rate?

(A) 920 kg/d
(B) 990 kg/d
(C) 1100 kg/d
(D) 1700 kg/d

69. What is the bed area when the GAC use rate is 1000 kg/d?

(A) 1.5 m²
(B) 4.2 m²
(C) 6.7 m²
(D) 15 m²

70. What is the head loss through the adsorber per 10 m² of bed area through for a Reynolds number of 1.0×10^3?

(A) 4.7 m
(B) 58 m
(C) 94 m
(D) 320 m

SITUATION FOR PROBLEM 71

A surface condenser is used to remove water vapor from an air stream. The operating temperatures for the condenser are as follows.

gas inlet temperature	198°F
gas outlet temperature	80°F
liquid inlet temperature	40°F
liquid outlet temperature	130°F

71. What is the average temperature change through the condenser?

(A) 28°F
(B) 53°F
(C) 110°F
(D) 160°F

SITUATION FOR PROBLEMS 72–73

The following problems address carbon monoxide in the ambient air.

72. Which statement inaccurately characterizes the health effects of carbon monoxide exposure?

(A) It is lethal at concentrations exceeding 5000 ppm.
(B) It heightens nervousness and mental agitation.
(C) It deprives the body of oxygen.
(D) It forms carboxyhemoglobin with blood.

73. Which statement inaccurately characterizes the physical characteristics of carbon monoxide?

(A) It is a colorless and odorless gas.
(B) It reacts to form carbon dioxide.
(C) It is formed by incomplete combustion.
(D) It accumulates in the atmosphere.

SITUATION FOR PROBLEMS 74–75

The maximum concentrations of air pollutants measured in a metropolitan area during a single day are as follows.

pollutant	duration (h)	concentration
O_3	1	0.21 ppm
CO	8	11 ppm
PM_{10}	24	240 µg/m³
SO_2	24	0.33 ppm
NO_2	1	0.8 ppm

74. What is the pollutant standard index (PSI) for the day?

(A) 133
(B) 160
(C) 185
(D) 233

75. What is the air quality descriptor for a pollutant standard index of 230?

(A) moderate
(B) unhealthful
(C) very unhealthful
(D) hazardous

SITUATION FOR PROBLEMS 76–77

A wind rose depicting wind data for a particular geographic location is shown in the following illustration.

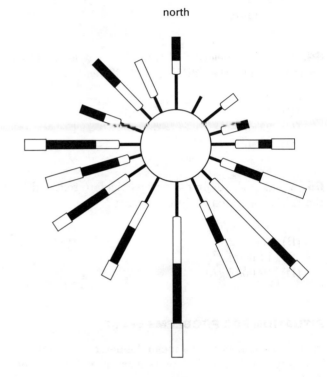

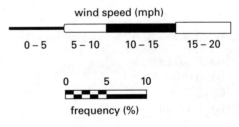

76. Based on the illustration, from which direction does the prevailing wind originate?

(A) from the north
(B) from the south
(C) from the north-northeast
(D) from the south-southwest

77. In the geographic location covered by the illustration, approximately what percent of the time does the wind speed blowing from W-SW exceed 10 mph?

(A) 2%
(B) 5%
(C) 8%
(D) 10%

SITUATION FOR PROBLEMS 78–82

The following problems pertain to risk assessment associated with exposure of human populations to toxic chemicals.

78. What is the primary difference in the dose-response relationship when comparing carcinogens with noncarcinogens?

(A) The slope of the dose-response curve is steeper for carcinogens than for noncarcinogens.
(B) The response from exposure to carcinogens is constant regardless of dose.
(C) A threshold dose exists for noncarcinogens below which no response is expected; there is no threshold for carcinogens.
(D) There are no general differences between the two.

79. What is the potency factor of a chemical?

(A) It is the dose of a toxic chemical that results in a risk of 1 in 1 million.
(B) It is the slope of the linear portion of the dose-response curve.
(C) It is the threshold concentration from the dose-response curve.
(D) It is the concentration at which any response is observed in half of an exposed population of laboratory animals.

80. What is the likelihood that a group A compound is carcinogenic to humans?

(A) Unlikely, because group A compounds show no evidence of carcinogenicity.
(B) Possible, although group A compounds show limited evidence of carcinogenicity in animals and no evidence of carcinogenicity in humans.
(C) Probable, although limited evidence exists showing group A compounds to be carcinogenic to humans.
(D) Likely, because sufficient evidence exists to show an association with cancer and exposure to a group A compound.

81. What is the level of exposure to a toxic chemical that a population can be assumed to endure without appreciable risk resulting?

(A) lowest observed adverse effect level (LOAEL)
(B) lowest observed effect level (LOEL)
(C) no observed effect level (NOEL)
(D) reference dose (RfD)

82. What parameter is used to evaluate the bioconcentration of a toxic contaminant in fish tissue?

(A) chronic daily intake (CDI)
(B) reference dose (RfD)
(C) bioconcentration factor (BCF)
(D) acceptable daily intake (ADI)

SITUATION FOR PROBLEMS 83–86

A workplace exposure survey has produced the following results.

	acetone	benzene	sec-butanol
1 h exposure at concentration (ppm)	1080	–	180
2 h exposure at concentration (ppm)	630	–	85
3 h exposure at concentration (ppm)	–	12	–
5 h exposure at concentration (ppm)	470	4	15
8 h TWA PEL* (ppm)	1000	10	150
acceptable ceiling (ppm)	–	25	–
acceptable maximum peak above ceiling concentration (ppm)	–	50	–
duration (min)	–	10	–

*TWA PEL is the time-weighted average peak exposure limit.

83. What is the cumulative exposure to benzene?

(A) 2 ppm
(B) 3 ppm
(C) 7 ppm
(D) 8 ppm

84. What is the cumulative exposure to the acetone and sec-butanol mixture?

(A) 0.56
(B) 0.94
(C) 1.8
(D) 640

85. If the cumulative exposure to the acetone, benzene, and sec-butanol mixture is 1.6, is it acceptable?

(A) Yes, because the cumulative exposure for the mixture is less than 1.0.
(B) Yes, because the cumulative exposure for the mixture is greater than 1.0.
(C) No, because the cumulative exposure for the mixture is less than 1.0.
(D) No, because the cumulative exposure for the mixture is greater than 1.0.

86. If a peak benzene exposure of 42 ppm occurs for 8 min during the first hour, is the cumulative exposure to benzene acceptable?

(A) Yes, because the cumulative exposure for benzene is greater than 1.0.
(B) No, because the cumulative exposure for benzene is less than 1.0.
(C) Yes, because the cumulative exposure for benzene is less than the TWA PEL for benzene.
(D) No, because the cumulative exposure for benzene is greater than the TWA PEL for benzene.

SITUATION FOR PROBLEMS 87–89

The following problems relate to radioactive materials and radiation exposure.

87. Which of the following federal statutes controls radiation exposure from X rays and other medical uses associated with radiation?

(A) Atomic Energy Act of 1946 and its amendments
(B) Low-Level Waste Policy Act of 1980 and its amendments
(C) Occupational Safety and Health Act of 1970
(D) Radiation Control for Health and Safety Act of 1968

88. For an absorbed alpha radiation dose of 0.13 rad, what is the dose equivalent?

(A) 0.13 rem
(B) 0.65 rem
(C) 1.3 rem
(D) 2.6 rem

89. An instrument used to measure gamma radiation detects 0.016 R/min at a distance from the source of 2 m. What should the instrument measure at 10 m from the source if the inverse-square law is used?

(A) 0.010 mR/min
(B) 0.64 mR/min
(C) 3.2 mR/min
(D) 7.2 mR/min

SITUATION FOR PROBLEMS 90–94

An underground storage tank piping failure has resulted in the release of 1500 gal of unleaded gasoline into the surrounding soil. Site conditions will not allow excavation of the tank or excavation of the contaminated soil. A soil vapor extraction system has been constructed as the remediation alternative. The gasoline contains the following chemicals of concern.

	chemical			
	A	B	C	D
weight fraction in gasoline (g/g)	0.018	0.082	0.021	0.029
molecular weight (g/mol)	78	92	106	106
mole fraction	0.020	0.091	0.018	0.016
vapor pressure (atm)	0.10	0.029	0.0092	0.0076
Henry's constant (atm·m^3/mol)	0.0055	0.0064	0.0084	0.0069
MCL (μg/L)	5	1000	700	10 000

The specific gravity of gasoline is approximately 0.72. The extraction system was designed to provide an air flow rate of 20 ft^3/min. The ambient soil temperature is 8°C.

90. Which of the four chemicals of concern will likely control design?

(A) chemical A
(B) chemical B
(C) chemical C
(D) chemical D

For Probs. 91 - 92, the source is nondiminishing and the vented chemical vapor is emitted directly to the atmosphere without treatment.

91. What is the mass of chemical A released to the soil?

(A) 19 kg
(B) 74 kg
(C) 100 kg
(D) 140 kg

92. What is the mass emission rate of chemical C to the atmosphere?

(A) 0.020 kg/d
(B) 0.25 kg/d
(C) 0.55 kg/d
(D) 0.62 kg/d

For Probs. 93 and 94, the source is diminishing, 90% of the chemical is extracted over an 8 month period, and 95% of the extracted vapor is destroyed before being emitted to the atmosphere.

93. What is the total mass of chemical B extracted from the soil?

 (A) 17 kg
 (B) 64 kg
 (C) 140 kg
 (D) 300 kg

94. What is the average concentration of chemical D emitted to the atmosphere?

 (A) 0.027 g/m^3
 (B) 0.20 g/m^3
 (C) 11 g/m^3
 (D) 26 g/m^3

SITUATION FOR PROBLEMS 95–96

The following problems pertain to radon in drinking water.

95. What is the primary exposure route for radon in drinking water?

 (A) Radon gas is emitted from groundwater when used in showers, bathing, and other in-home activities.
 (B) Radon gas is emitted from surface water when used in showers, bathing, and other in-home activities.
 (C) Particulate radon is ingested by drinking surface water.
 (D) Radon is ingested from eating foods grown in radon-containing soil.

96. What are the preferred mitigation measures for radon contamination in water supplies?

 (A) aeration
 (B) granular activated carbon adsorption
 (C) both (A) and (B)
 (D) neither (A) nor (B)

SITUATION FOR PROBLEMS 97–98

The following problems relate to indoor air pollution.

97. What contaminants associated with indoor air quality are generally considered to be the most important from a human health perspective?

 (A) asbestos and formaldehyde
 (B) environmental tobacco smoke and asbestos
 (C) radon and asbestos
 (D) radon and environmental tobacco smoke

98. What diseases are typically associated with non-occupational exposure to asbestos fibers?

 (A) asbestosis and cystic fibrosis
 (B) asbestosis and lung cancer
 (C) black lung and mesothelioma
 (D) lung cancer and mesothelioma

SITUATION FOR PROBLEMS 99–100

A public water supply is disinfected with chlorine to inactivate giardia cysts. The chlorine feed occurs as the water enters the wet well where it experiences a minimum hydraulic residence time of 30 min. The water is at 5°C.

99. What is the free chlorine concentration required to achieve a 4-log reduction in giardia cysts for a water pH of 7.7?

 (A) 5.7 mg/L
 (B) 7.3 mg/L
 (C) 9.1 mg/L
 (D) 11 mg/L

100. What is the minimum free chloride concentration within the optimum water pH range required to achieve a 4-log reduction in giardia cysts?

 (A) 6.2 mg/L
 (B) 7.3 mg/L
 (C) 9.1 mg/L
 (D) 10 mg/L

Exam 1—Solutions

1. C
2. B
3. B
4. C
5. B
6. C
7. B
8. A
9. C
10. A
11. B
12. C
13. B
14. C
15. B
16. D
17. B
18. B
19. C
20. A
21. A
22. C
23. B
24. C
25. C
26. B
27. C
28. B
29. D
30. A
31. D
32. B
33. D
34. A
35. C
36. C
37. D
38. C
39. B
40. D
41. C
42. D
43. B
44. B
45. A
46. B
47. B
48. A
49. C
50. A
51. C
52. A
53. C
54. C
55. A
56. D
57. D
58. A
59. A
60. C
61. B
62. B
63. A
64. C
65. D
66. B
67. C
68. B
69. B
70. C
71. B
72. B
73. D
74. D
75. C
76. B
77. C
78. C
79. B
80. D
81. D
82. C
83. C
84. B
85. D
86. C
87. D
88. D
89. B
90. A
91. B
92. D
93. D
94. A
95. A
96. C
97. D
98. D
99. D
100. A

INFORMATION FOR SOLUTIONS 1–4

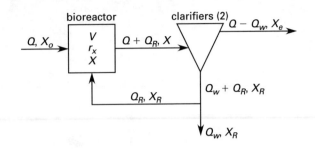

Q	influent flow from primary clarifiers	$60\,000$ m^3/d
Q_R	return solids flow rate	m^3/d
Q_w	wasted solids flow rate	m^3/d
r_x	biomass production rate	0.5 kg/m$^3 \cdot$d
V	bioreactor volume	8000 m^3
X	mixed liquor VSS (MLVSS)	mg/L
X_e	effluent VSS from secondary clarifiers	60 mg/L
X_o	influent VSS from primary clarifiers	100 mg/L
X_R	return flow VSS	12 000 mg/L

1. Perform a solids mass balance around the bioreactor.

Write the statement in words.

$$\text{influent VSS} + \text{return VSS} + \text{cell growth} = \text{effluent VSS}$$

Write the equation with variables.

$$QX_o + Q_R X_R + V r_x = (Q + Q_R)X$$

Write the equation with values.

$$\left(60\,000\ \frac{\text{m}^3}{\text{d}}\right)\left(100\ \frac{\text{mg}}{\text{L}}\right)\left(10^{-3}\ \frac{\text{kg}\cdot\text{L}}{\text{mg}\cdot\text{m}^3}\right)$$
$$+ Q_R \left(12\,000\ \frac{\text{mg}}{\text{L}}\right)\left(10^{-3}\ \frac{\text{kg}\cdot\text{L}}{\text{mg}\cdot\text{m}^3}\right)$$
$$+ (8000\ \text{m}^3)\left(0.5\ \frac{\text{kg}}{\text{m}^3\cdot\text{d}}\right)$$
$$= \left(60\,000\ \frac{\text{m}^3}{\text{d}} + Q_R\right)\left(1800\ \frac{\text{mg}}{\text{L}}\right)$$
$$\times \left(10^{-3}\ \frac{\text{kg}\cdot\text{L}}{\text{mg}\cdot\text{m}^3}\right)$$

Multiply through and combine terms.

$$6000\ \frac{\text{kg}}{\text{d}} + \left(12\ \frac{\text{kg}}{\text{m}^3}\right)Q_R + 4000\ \frac{\text{kg}}{\text{d}}$$
$$= 108\,000\ \frac{\text{kg}}{\text{d}} + \left(1.8\ \frac{\text{kg}}{\text{m}^3}\right)Q_R$$

Rearrange and solve for Q_R.

$$108\,000\ \frac{\text{kg}}{\text{d}} - 6000\ \frac{\text{kg}}{\text{d}} - 4000\ \frac{\text{kg}}{\text{d}}$$
$$= \left(12\ \frac{\text{kg}}{\text{m}^3} - 1.8\ \frac{\text{kg}}{\text{m}^3}\right)Q_R$$

$$Q_R = \frac{98\,000\ \frac{\text{kg}}{\text{d}}}{10.2\ \frac{\text{kg}}{\text{m}^3}}$$
$$= 9608\ \text{m}^3/\text{d}\quad (9600\ \text{m}^3/\text{d})$$

The answer is (C).

2. Perform a solids mass balance around the secondary clarifiers.

Write the statement in words.

$$\text{influent VSS} = \text{effluent VSS} + \text{return VSS} + \text{wasted VSS}$$

Write the equation with variables.

$$(Q + Q_R)X = (Q - Q_w)X_e + (Q_w + Q_R)X_R$$

Write the equation with values.

$$\left(60\,000\ \frac{\text{m}^3}{\text{d}} + 5000\ \frac{\text{m}^3}{\text{d}}\right)\left(2200\ \frac{\text{mg}}{\text{L}}\right)\left(10^{-3}\ \frac{\text{kg}\cdot\text{L}}{\text{mg}\cdot\text{m}^3}\right)$$
$$= \left(60\,000\ \frac{\text{m}^3}{\text{d}} - Q_w\right)\left(60\ \frac{\text{mg}}{\text{L}}\right)\left(10^{-3}\ \frac{\text{kg}\cdot\text{L}}{\text{mg}\cdot\text{m}^3}\right)$$
$$+ \left(Q_w + 5000\ \frac{\text{m}^3}{\text{d}}\right)\left(12000\ \frac{\text{mg}}{\text{L}}\right)$$
$$\times \left(10^{-3}\ \frac{\text{kg}\cdot\text{L}}{\text{mg}\cdot\text{m}^3}\right)$$

Multiply through and combine terms.

$$143\,000\ \frac{\text{kg}}{\text{d}} = 3600\ \frac{\text{kg}}{\text{d}} - \left(0.060\ \frac{\text{kg}}{\text{m}^3}\right)Q_w$$
$$+ \left(12\ \frac{\text{kg}}{\text{m}^3}\right)Q_w + 60\,000\ \frac{\text{kg}}{\text{d}}$$

Rearrange and solve for Q_w.

$$Q_w = \frac{143\,000\ \frac{\text{kg}}{\text{d}} - 3600\ \frac{\text{kg}}{\text{d}} - 60000\ \frac{\text{kg}}{\text{d}}}{12\ \frac{\text{kg}}{\text{m}^3} - 0.60\ \frac{\text{kg}}{\text{m}^3}}$$
$$= 6965\ \text{m}^3/\text{d}\quad [\text{for both clarifiers combined}]$$

$$\frac{Q_w}{\text{clarifier}} = \frac{6965 \, \frac{\text{m}^3}{\text{d}}}{2 \text{ clarifiers}}$$
$$= 3482 \text{ m}^3/\text{d·clarifier} \quad (3500 \text{ m}^3/\text{d·clarifier})$$

The answer is (B).

3. Calculate the mass of dry solids wasted per day.

$$\left(1000 \, \frac{\text{m}^3}{\text{d}}\right)\left(12\,000 \, \frac{\text{mg}}{\text{L}}\right)\left(10^{-6} \, \frac{\text{kg}}{\text{mg}}\right)$$
$$\times \left(10^3 \, \frac{\text{L}}{\text{m}^3}\right) = 12\,000 \text{ kg/d}$$

Since the specific gravity of the volatile solids is 1.0, the density of the wasted solids is the same as that of water, 1000 kg/m³.

The volume of wasted solids at 20% is

$$\frac{12\,000 \, \frac{\text{kg}}{\text{d}}}{\left(1000 \, \frac{\text{kg}}{\text{m}^3}\right)(0.20)} = 60 \text{ m}^3/\text{d}$$

The answer is (B).

4. To keep the MLVSS constant, the solids wasting rate must equal the biomass production rate.

$$\left(0.5 \, \frac{\text{kg}}{\text{m}^3 \cdot \text{d}}\right)(8000 \text{ m}^3) = 4000 \text{ kg/d}$$

The answer is (C).

5.

pipe no.	pipe diameter, D (mm)	pipe length, L (km)	average flow, Q_{ave} (L/d)	average flow, Q'_{ave} (L/km·mm·d)
1	600	0.085	2500	49.0
2	600	0.046	2000	72.5
3	300	0.040	140	11.7
4	250	0.067	500	29.8
5	200	0.091	50	2.75
6	200	0.131	80	3.10
7	150	0.050	110	14.7

$$Q'_{ave} = \frac{Q_{ave}}{DL}$$

Typical specification for infiltration into new sewer systems is 45 L/km·mm·d. Pipes 1 and 2 exceed the typical value.

The answer is (B).

6. Refer to the table of values for Sol. 6.

$$Q'_{ave} = \frac{Q_{ave}}{DL}$$

$$\text{percent of I/I based on L/km·mm·d} = \frac{Q'_{ave} \times 100\%}{183.55 \text{ L/km·mm·d}}$$

Pipes 1, 2, and 4 account for 82.4% of the measured I/I.

The answer is (C).

7. The BOD₅ loading at average BOD₅ = 135 mg/L is

$$\left(135 \, \frac{\text{mg}}{\text{L}}\right)\left(50\,000 \, \frac{\text{m}^3}{\text{d}}\right)\left(10^{-3} \, \frac{\text{kg·L}}{\text{mg·m}^3}\right)$$
$$= 6750 \text{ kg/d}$$

Table for Sol. 6

pipe no.	pipe diameter, D (mm)	pipe length, L (km)	average flow, Q_{ave} (L/d)	average flow, Q'_{ave} (L/km·mm·d)	percent of I/I based on L/km·mm·d
1	600	0.085	2500	49.0	26.7
2	600	0.046	2000	72.5	39.5
3	300	0.040	140	11.7	6.4
4	250	0.067	500	29.8	16.2
5	200	0.091	50	2.75	1.5
6	200	0.131	80	3.10	1.7
7	150	0.050	110	14.7	8.0
			total average flow	183.55	

The BOD$_5$ loading from domestic sources is

$$6750 \frac{\text{kg}}{\text{d}} - \left(348 \frac{\text{mg}}{\text{L}}\right)\left(8000 \frac{\text{m}^3}{\text{d}}\right)\left(10^{-3} \frac{\text{kg} \cdot \text{L}}{\text{mg} \cdot \text{m}^3}\right)$$
$$= 3966 \text{ kg/d}$$

The domestic flow rate based on typical BOD$_5$ = 200 mg/L is

$$\frac{\left(3966 \frac{\text{kg}}{\text{d}}\right)\left(10^3 \frac{\text{mg} \cdot \text{m}^3}{\text{kg} \cdot \text{L}}\right)}{200 \frac{\text{mg}}{\text{L}}} = 19\,830 \text{ m}^3/\text{d}$$

The approximate I/I to the entire sewer system is

$$50\,000 \frac{\text{m}^3}{\text{d}} - 19\,830 \frac{\text{m}^3}{\text{d}} - 8000 \frac{\text{m}^3}{\text{d}}$$
$$= 22\,170 \text{ m}^3/\text{d} \quad (22\,000 \text{ m}^3/\text{d})$$

The answer is (B).

8. The typical average daily flow during the maximum month is 125% of the average daily flow during the year.

$$\frac{\left(9{,}000{,}000 \frac{\text{gal}}{\text{yr}}\right)\left(\frac{125\%}{100\%}\right)}{\left(12 \frac{\text{mo}}{\text{yr}}\right)\left(30 \frac{\text{day}}{\text{mo}}\right)}$$
$$= 31{,}250 \text{ gal/day during peak month}$$

The answer is (A).

9. Assume that the medium-strength wastewater average BOD concentration is 220 mg/L. This is typical.

$$\left(9{,}000{,}000 \frac{\text{gal}}{\text{day}}\right)\left(220 \frac{\text{mg}}{\text{L}}\right)\left(3.785 \frac{\text{L}}{\text{gal}}\right)\left(10^{-6} \frac{\text{kg}}{\text{mg}}\right)$$
$$= 7494 \text{ kg BOD/yr} \quad (7500 \text{ kg BOD/yr})$$

The answer is (C).

10. The BOD loading rate will remain unchanged since the population and wastewater generation activities remain unchanged. If only the flow decreases, then the concentration will increase and loading will remain constant.

The answer is (A).

11. Approximate detention basin routing for a type II storm is shown in the following illustration.

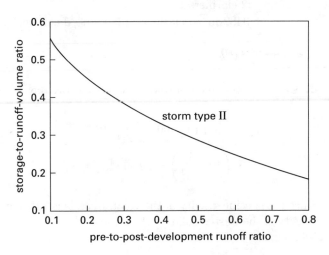

From the illustration, for a pre-to-post-development runoff ratio, Q_o/Q_i, of 0.42, the required storage-to-runoff volume ratio is

$$\frac{V_S}{V_R} = 0.32$$

The answer is (B).

12. Refer to the following illustration for Sol. 12. The runoff volume, V_R, for a 25 yr storm is

CN = 65 [woodlands]
CN = 98 [rooftops/pavement]
CN = 69 [landscaping]

$V_R = (0.2 \text{ in})\left(2.54 \frac{\text{cm}}{\text{in}}\right)$
$\quad = 0.51 \text{ cm}$ [woodlands] $\quad\quad$ 77 ha

$V_R = (1.8 \text{ in})\left(2.54 \frac{\text{cm}}{\text{in}}\right)$
$\quad = 4.6 \text{ cm}$ [rooftops/pavement] $\quad$ 33 ha

$V_R = (0.3 \text{ in})\left(2.54 \frac{\text{cm}}{\text{in}}\right)$
$\quad = 0.76 \text{ cm}$ [landscaping] $\quad\quad$ 82 ha
$\quad\quad\quad\quad\quad\quad\quad\quad\quad\quad\quad\quad\overline{192 \text{ ha}}$

$$V_{R,\text{ave}} = \frac{\begin{array}{c}(0.51 \text{ cm})(77 \text{ ha}) + (4.6 \text{ cm})\\ \times (33 \text{ ha}) + (0.76 \text{ cm})(82 \text{ ha})\end{array}}{192 \text{ ha}}$$
$$= 1.3 \text{ cm}$$

The answer is (C).

13. The required storage volume is

$$V_S = \left(\frac{V_S}{V_R}\right) V_R = (0.25)(1.2 \text{ cm})$$
$$= 0.30 \text{ cm}$$

The answer is (B).

Illustration for Sol. 12

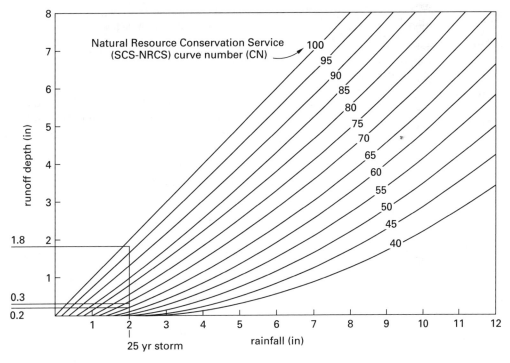

14. The basin capacity is

$$(0.95 \text{ cm})(192 \text{ ha})\left(\frac{1 \text{ m}}{100 \text{ cm}}\right)\left(10\,000 \,\frac{\text{m}^2}{\text{ha}}\right)$$
$$= 18\,240 \text{ m}^3 \quad (18\,000 \text{ m}^3)$$

The answer is (C).

15.

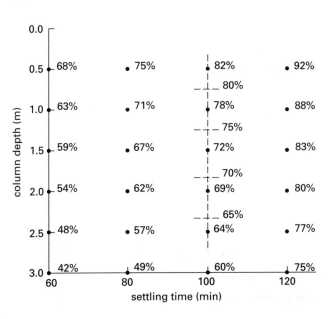

From the graph at $Z_o = 2.5$ m and $t = 100$ min, $h_o = 64\%$.

i	Z_i (m)	Δh	$Z_i \Delta h$
1	2.33	0.01	0.0233
2	1.83	0.05	0.0915
3	1.25	0.05	0.0625
4	0.76	0.05	0.0380
			0.22

Δh is the incremental fractional efficiency at each depth, Z_i.

The overall efficiency is

$$h_o + \frac{\sum (Z_i \Delta h)}{Z_o} = 0.64 + \frac{0.22}{2.5}$$
$$= 0.73 \quad (73\%)$$

The answer is (B).

16. The settling zone surface area per basin is

$$A_s = \frac{Qt}{Z_o(\text{number of basins})}$$
$$= \frac{\left(12\,000 \,\frac{\text{m}^3}{\text{d}}\right)(100 \text{ min})}{(2.5 \text{ m})\left(1440 \,\frac{\text{min}}{\text{d}}\right)(4 \text{ basins})}$$
$$= 83.2 \text{ m}^2/\text{basin}$$

The overflow rate is

$$q_o = \dfrac{Q}{\dfrac{\text{basin}}{A_s}} = \dfrac{\left(12\,000\ \dfrac{\text{m}^3}{\text{d}}\right)\left(1\ \dfrac{\text{d}}{24\ \text{h}}\right)}{\left(83.2\ \dfrac{\text{m}^2}{\text{basin}}\right)(4)}$$

$$= 1.5\ \text{m}^3/\text{m}^2\cdot\text{h}$$

The answer is (D).

17. From the graph at $Z_o = 1.5$ m and $t = 100$ min, $h_o = 72\%$.

i	Z_i (m)	Δh	$Z_i \Delta h$
1	1.25	0.03	0.0375
2	0.76	0.05	0.0380
			0.0755

The overall efficiency is

$$0.72 + \dfrac{0.0755}{1.5} = 0.77$$
$$77\% > 73\%$$

The efficiency increases.

The answer is (B).

18. From the graph at a settling zone depth of $Z_o = 2.5$ m and a settling time of $t = 120$ min, $h_o = 77\%$.

Because $h_o = 77\% > h_o = 64\%$ when $t = 100$ min, the overall efficiency will be greater than 73%.

The efficiency increases.

The answer is (B).

19. Assume at 18% solids, the settled solids density is the same as water, 1000 kg/m³.

The volume of TSS removed at 18% solids is

$$\dfrac{(0.80)\left(12\,000\ \dfrac{\text{m}^3}{\text{d}}\right)\left(195\ \dfrac{\text{mg}}{\text{L}}\right)\left(10^{-6}\ \dfrac{\text{kg}}{\text{mg}}\right)}{\left(1000\ \dfrac{\text{kg}}{\text{m}^3}\right)(0.18)\left(\dfrac{1\ \text{m}^3}{10^3\ \text{L}}\right)}$$

$$= 10.4\ \text{m}^3 \quad (10\ \text{m}^3)$$

The answer is (C).

INFORMATION FOR SOLUTIONS 20–22

The following table illustrates the relationships among various hydraulic residence times for ideal plug flow and ideal complete mix reactors.

plug flow	complete mix
modal time = theoretical time	modal time = 0
median time approaches theoretical time	median time approaches 0
minimum time = theoretical time	minimum time = 0

20. From the illustration, the modal and median hydraulic residence times are both approximately 40 min, and the minimum hydraulic residence time is approximately 20 min. A comparison of these approximate values to the relationship defined in the table indicates that the reactor more closely models plug flow than complete mix.

The answer is (A).

21. For a slug tracer in an ideal plug flow reactor,

$$\begin{array}{c}\text{modal hydraulic} \\ \text{residence time}\end{array} = \begin{array}{c}\text{theoretical hydraulic} \\ \text{residence time}\end{array}$$

The answer is (A).

22. For a slug tracer in an ideal complete mix reactor,

$$\text{modal hydraulic residence time} = 0$$

The answer is (C).

23. The air flow rate is

$$Q_a = Q\left(\dfrac{V_a}{V_w}\right) = Q\left(\dfrac{S}{K_H}\right)$$
$$= \left(375\ \dfrac{\text{gal}}{\text{min}}\right)\left(\dfrac{3.5}{0.18}\right)\left(3.785\ \dfrac{\text{L}}{\text{gal}}\right)\left(\dfrac{1\ \text{min}}{60\ \text{s}}\right)$$
$$= 460\ \text{L/s}$$

The answer is (B).

24.

$$\dfrac{C_0}{C} = \dfrac{2000\ \dfrac{\mu\text{g}}{\text{L}}}{20\ \dfrac{\mu\text{g}}{\text{L}}} = 100$$

$$\text{NTU} = \left(\dfrac{S}{S-1}\right)\ln\left(\dfrac{\left(\dfrac{C_0}{C}\right)(S-1)+1}{S}\right)$$

$$= \left(\dfrac{3.5}{3.5-1}\right)\ln\left(\dfrac{(100)(3.5-1)+1}{3.5}\right)$$

$$= 6.0$$

The answer is (C).

25.

$$\text{HTU} = \frac{\text{HLR}_m}{K_{La}\rho_w}$$

ρ_w water density   1000 kg/m³
HLR_m mass hydraulic loading rate   kg/m²s

$$\text{HLR}_m = \left(30 \ \frac{\text{gal}}{\text{min-ft}^2}\right)\left(3.785 \ \frac{\text{L}}{\text{gal}}\right)\left(10.76 \ \frac{\text{ft}^2}{\text{m}^2}\right)$$
$$\times \left(\frac{1 \ \text{min}}{60 \ \text{s}}\right)\left(1000 \ \frac{\text{kg}}{\text{m}^3}\right)\left(\frac{1 \ \text{m}^3}{1000 \ \text{L}}\right)$$
$$= 20 \ \text{kg/m}^2\text{s}$$

$$\text{HTU} = \frac{20 \ \frac{\text{kg}}{\text{m}^2\text{s}}}{\left(\frac{0.014}{\text{s}}\right)\left(1000 \ \frac{\text{kg}}{\text{m}^3}\right)}$$
$$= 1.43 \ \text{m} \quad (1.4 \ \text{m})$$

The answer is (C).

26.
The column area is

$$A = \frac{Q}{\text{HLR}} = \frac{375 \ \frac{\text{gal}}{\text{min}}}{30 \ \frac{\text{gal}}{\text{min-ft}^2}}$$
$$= 12.5 \ \text{ft}^2$$

The column diameter is

$$d = \frac{\sqrt{(12.5 \ \text{ft}^2)\left(\frac{4}{\pi}\right)}}{3.28 \ \frac{\text{ft}}{\text{m}}}$$
$$= 1.2 \ \text{m}$$

The answer is (B).

27.

$$\frac{Q}{\text{tank}} = \frac{7500 \ \frac{\text{m}^3}{\text{d}}}{2 \ \text{tanks}}$$
$$= 3750 \ \text{m}^3/\text{d·tank}$$
$$q_o = \frac{Q}{A_s}$$

The settling zone surface area is

$$A_s = \frac{Q}{q_o} = \frac{3750 \ \frac{\text{m}^3}{\text{d·tank}}}{\left(1.8 \ \frac{\text{m}^3}{\text{m}^2\cdot\text{h}}\right)\left(24 \ \frac{\text{h}}{\text{d}}\right)}$$
$$= 86.8 \ \text{m}^2/\text{tank} \quad (87 \ \text{m}^2/\text{tank})$$

The answer is (C).

28.

$$\frac{Q}{\text{tank}} = \frac{7500 \ \frac{\text{m}^3}{\text{d}}}{2 \ \text{tanks}} = 3750 \ \frac{\text{m}^3}{\text{d·tank}}$$

$$q_w = \frac{Q}{L_w} = \frac{3750 \ \frac{\text{m}^3}{\text{d·tank}}}{\left(14 \ \frac{\text{m}^3}{\text{m·h}}\right)\left(24 \ \frac{\text{h}}{\text{d}}\right)}$$
$$= 11.2 \ \text{m/tank} \quad (11 \ \text{m/tank})$$

The answer is (B).

29.
Algae problems are associated with nutrients and stagnant water. The stagnant water allows the sunlight required by the algae to penetrate below the water surface. Baffles would improve hydraulic efficiency but would leave the water surface unbroken and leave the algae exposed to the sunlight. The same effect would be observed from discontinued surface aeration, which would actually promote algae production, since aeration agitates the surface and interferes with the ability of the algae to use the sunlight.

Algae congregate near the surface to maximize their exposure to sunlight, so an effluent structure that takes the discharge from near the surface will skim algae and increase, rather than decrease, TSS discharges. A better alternative would be to take the discharge from some distance below the surface.

For these reasons, none of the choices provided would contribute to controlling algae associated with TSS discharges from pond 3.

The answer is (D).

30.
Assuming carbon is represented by BOD, typical BOD:N:P for bacterial cells is 60:12:1. 200:15:4 is the BOD:N:P for the pond influent. Let subscript c represent cells and subscript o represent influent.

The nitrogen requirement is

$$\frac{\text{BOD}_o}{\text{BOD}_c} = \frac{N_o}{N_c}$$

$$\frac{200 \ \frac{\text{mg}}{\text{L}}}{60 \ \frac{\text{mg}}{\text{L}}} = \frac{N_o}{12 \ \frac{\text{mg}}{\text{L}}}$$

$$N_o = 40 \ \frac{\text{mg}}{\text{L}} > 15 \ \frac{\text{mg}}{\text{L}}$$

Therefore, the influent is likely to be deficient in nitrogen.

The phosphorous requirement is

$$\frac{\text{BOD}_o}{\text{BOD}_c} = \frac{P_o}{P_c}$$

$$\frac{200 \ \frac{mg}{L}}{60 \ \frac{mg}{L}} = \frac{P_o}{1 \ \frac{mg}{L}}$$

$$P_o = 3.33 \ \frac{mg}{L} < 4 \ \frac{mg}{L}$$

Therefore, the influent likely contains adequate amounts of phosphorous.

The answer is (A).

31. The hydraulic efficiency of pond 2 could be increased by installing surface aerators to completely mix the pond's contents, by relocating inlet and outlet structures to provide maximum separation distance, and/or by installing baffles to increase the hydraulic path.

The answer is (D).

32.

$$\text{discharge BOD}_u = \frac{\text{BOD}_t}{(1 - e^{-k_d t})}$$

$$= \frac{\left(198 \ \frac{mg}{L}\right)}{(1 - e^{-(0.26/d)(5 \ d)})}$$

$$= 272 \ mg/L$$

Assume stream deoxygenation constant, k_d, applies to BOD.

$$\text{river BOD}_u = \frac{2.1 \ \frac{mg}{L}}{(1 - e^{-(0.16/d)(5d)})}$$

$$= 3.8 \ mg/L$$

Use a mass balance to find the BOD of the mixed flows.

$$\text{BOD}_u \text{ of mixed flows} = \frac{\left(2.65 \ \frac{ft^3}{s}\right)\left(272 \ \frac{mg}{L}\right) + \left(68 \ \frac{ft^3}{s}\right)\left(3.8 \ \frac{mg}{L}\right)}{2.65 \ \frac{ft^3}{s} + 68 \ \frac{ft^3}{s}}$$

$$= 13.86 \ mg/L \quad (14 \ mg/L)$$

The answer is (B).

33. The saturated dissolved oxygen in the river above the discharge is

$$\text{DO}_{sat} = 11.59 \ mg/L \text{ at } 48°F$$

The dissolved oxygen deficit at the discharge is

$$D_o = \text{DO}_{sat} - \text{DO} = 11.59 \ \frac{mg}{L} - 8.9 \ \frac{mg}{L}$$

$$= 2.69 \ mg/L$$

$$a = -k_d t = -\left(\frac{0.16}{d}\right)(14 \ d)$$

$$= -2.24$$

$$b = -k_r t = -\left(\frac{0.23}{d}\right)(14 \ d)$$

$$= -3.22$$

The DO concentration is

$$D = \frac{k_d \text{BOD}_u (e^a - e^b)}{k_r - k_d} + D_o e^b$$

$$= \frac{\left(\frac{0.16}{d}\right)\left(10 \ \frac{mg}{L}\right)(e^{-2.24} - e^{-3.22})}{\frac{0.23}{d} - \frac{0.16}{d}}$$

$$+ \left(2.69 \ \frac{mg}{L}\right)(e^{-3.22})$$

$$= 1.63 \ \frac{mg}{L}$$

$$\text{DO} = 11.59 \ \frac{mg}{L} - 1.63 \ \frac{mg}{L}$$

$$= 9.96 \ mg/L \quad (10 \ mg/L)$$

The answer is (D).

34. The saturated dissolved oxygen in the river above the discharge is

$$\text{DO}_{sat} = 11.59 \ mg/L \text{ at } 48°F$$

The dissolved oxygen deficit at the discharge is

$$D_o = \text{DO}_{sat} - \text{DO} = 11.59 \ \frac{mg}{L} - 8.9 \ \frac{mg}{L}$$

$$= 2.69 \ \frac{mg}{L}$$

$$a = -k_d t_c = \left(\frac{-0.16}{d}\right)(3.8 \ d)$$

$$= -0.608$$

$$b = -k_r t_c = \left(\frac{-0.23}{d}\right)(3.8 \ d)$$

$$= -0.874$$

The critical DO deficit is

$$D_c = \frac{k_d \text{BOD}_u (e^a - e^b)}{k_r - k_d} + D_o e^b$$

$$= \frac{\left(\frac{0.16}{d}\right)\left(10 \ \frac{mg}{L}\right)(e^{-0.608} - e^{-0.874})}{\frac{0.23}{d} - \frac{0.16}{d}}$$

$$+ \left(2.69 \ \frac{mg}{L}\right)(e^{-0.874})$$

$$= 4.03 \ mg/L \quad (4.0 \ mg/L)$$

The answer is (A).

35. The saturated dissolved oxygen in the river above the discharge is

$$DO_{sat} = 11.59 \text{ mg/L at } 48°F$$

The dissolved oxygen deficit at the discharge is

$$D_o = DO_{sat} - DO = 11.59 \frac{mg}{L} - 8.9 \frac{mg}{L}$$
$$= 2.69 \text{ mg/L}$$

The critical time is

$$t_c = (k_r - k_d)^{-1} \ln\left(\left(\frac{k_r}{k_d}\right)\left(1 - \frac{D_o(k_r - k_d)}{k_d BOD_u}\right)\right)$$

$$= \left(\frac{0.23}{d} - \frac{0.16}{d}\right)^{-1}$$

$$\times \ln\left(\left(\frac{\frac{0.23}{d}}{\frac{0.16}{d}}\right)\left(1 - \frac{\left(2.69 \frac{mg}{L}\right) \times \left(\frac{0.23}{d} - \frac{0.16}{d}\right)}{\left(\frac{0.16}{d}\right)\left(10 \frac{mg}{L}\right)}\right)\right)$$

$$= 3.40 \text{ d} \quad (3.4 \text{ d})$$

The answer is (C).

36. From the universal gas law, the molar volume of a gas at 20°C and 1 atm is 22.4 L/mol.

$$CH_4 \text{ MW} = 12 \frac{g}{mol} + (4)\left(1 \frac{g}{mol}\right) = 16 \frac{g}{mol}$$

$$CH_4 \text{ mass} = \frac{(0.70)\left(525{,}000 \frac{ft^3}{d}\right)\left(16 \frac{g}{mol}\right) \times \left(2.204 \frac{lbm}{kg}\right)}{\left(22.4 \frac{L}{mol}\right)\left(0.0353 \frac{ft^3}{L}\right)\left(1000 \frac{g}{kg}\right)}$$

$$= 16{,}390 \text{ lbm/d} \quad (16{,}000 \text{ lbm/day})$$

The answer is (C).

37. The combustion reaction is

$$CH_4 + 2O_2 \rightarrow CO_2 + 2H_2O$$

$\Delta H° CH_4$ -74.82 kJ/mol
$\Delta H° O_2$ 0 (standard state)
$\Delta H° CO_2$ -393.5 kJ/mol
$\Delta H° H_2O$ -241.8 kJ/mol

$$\Delta H°_{reaction} = \Delta H°_{products} - \Delta H°_{reactants}$$
$$= (1 \text{ mol})\left(-393.5 \frac{kJ}{mol}\right)$$
$$+ (2 \text{ mol})\left(-241.8 \frac{kJ}{mol}\right)$$
$$- (1 \text{ mol})\left(-74.82 \frac{kJ}{mol}\right)$$
$$- (2 \text{ mol})\left(0 \frac{kJ}{mol}\right)$$
$$= -802.3 \frac{kJ}{mol}$$

The reaction is exothermic.

The heating value per 1000 ft³ of CH_4 at 20°C and 1 atm is

$$\frac{(1000 \text{ ft}^3)\left(28.3 \frac{L}{ft^3}\right)\left(802.3 \frac{kJ}{mol}\right)}{22.4 \frac{L}{mol}}$$
$$= 1.0 \times 10^6 \text{ kJ}/1000 \text{ ft}^3 \text{ } CH_4$$

The answer is (D).

38.

c	specific heat of water	4.184 J/g°C
q	heat required	kJ
ρ	water density	1000 g/L
V	volume of water	25,000 gal
ΔT	temperature change	°C

$$\Delta T = 60°C - 10°C = 50°C$$
$$q = cV\Delta T\rho$$
$$= \left(4.184 \frac{J}{g°C}\right)(25{,}000 \text{ gal})(50°C)\left(3.785 \frac{L}{gal}\right)$$
$$\times \left(1000 \frac{g}{L}\right)\left(\frac{1 \text{ kJ}}{1000 \text{ J}}\right)$$
$$= 2.0 \times 10^7 \text{ kJ}$$

From the universal gas law, the molar volume of a gas at 20°C and 1 atm is 22.4 L/mol.

$$CH_4 \text{ MW} = 12 \frac{g}{mol} + (4)\left(1 \frac{g}{mol}\right)$$
$$= 16 \text{ g/mol}$$

$$CH_4 \text{ volume} = \frac{(2.0 \times 10^7 \text{ kJ})\left(22.4 \frac{L}{mol}\right) \times \left(0.0353 \frac{ft^3}{L}\right)}{800 \frac{kJ}{mol}}$$

$$= 19{,}770 \text{ ft}^3 \quad (20{,}000 \text{ ft}^3)$$

The answer is (C).

39. Hydrochloric acid is HCl and sodium hydroxide is NaOH.

$$HCl + NaOH \rightarrow H_2O + Na^+ + Cl^-$$

On the left side of the equation,
H: $(2 \text{ mol})\left(1 \dfrac{g}{mol}\right) = 2$ g
O: $(1 \text{ mol})\left(16 \dfrac{g}{mol}\right) = 16$ g
Cl: $(1 \text{ mol})\left(35.5 \dfrac{g}{mol}\right) = 35.5$ g
Na: $(1 \text{ mol})\left(23 \dfrac{g}{mol}\right) = 23$ g
$\overline{76.5 \text{ g}}$

net charge = 0

On the right side,
H: $(2 \text{ mol})\left(1 \dfrac{g}{mol}\right) = 2$ g
O: $(1 \text{ mol})\left(16 \dfrac{g}{mol}\right) = 16$ g
Cl: $(1 \text{ mol})\left(35.5 \dfrac{g}{mol}\right) = 35.5$ g
Na: $(1 \text{ mol})\left(23 \dfrac{g}{mol}\right) = 23$ g
$\overline{76.5 \text{ g}}$

net charge = 0

The equation is mass and charge balanced.

The answer is (B).

40. If a 1-to-1 molar ratio is assumed, 1 mol of sodium hydroxide (NaOH) will neutralize 1 mol of hydrochloric acid (HCl). Therefore, 2.1 mol of NaOH is needed per liter of 2.1 molar HCl treated.

$$\text{MW NaOH} = 23 \dfrac{g}{mol} + 16 \dfrac{g}{mol} + 1 \dfrac{g}{mol}$$
$$= 40 \text{ g/mol}$$

$$M_{NaOH} = \left(2.1 \dfrac{mol}{L}\right)\left(40 \dfrac{g}{mol}\right)\left(300\,000 \dfrac{L}{d}\right)$$
$$\times \left(\dfrac{1 \text{ kg}}{1000 \text{ g}}\right)$$
$$= 25\,200 \text{ kg/d} \quad (25\,000 \text{ kg/d})$$

The answer is (D).

41. The standard enthalpy for the reaction is

$$\Delta H°_{products} - \Delta H°_{reactants} = -690 \dfrac{kJ}{mol}$$
$$- \left(-640 \dfrac{kJ}{mol}\right)$$
$$= -50 \text{ kJ/mol}$$

The reaction is exothermic and the temperature increases.

$$\Delta T = \dfrac{q}{mc}$$

c	specific heat of water	4.184 J/g°C
m	mass of water	assume 1000 g/L
q	heat required	kJ
ΔT	temperature change	°C

$$\Delta T = \dfrac{\left(50 \dfrac{kJ}{mol}\right)\left(2.1 \dfrac{mol}{L}\right)\left(1000 \dfrac{J}{kJ}\right)}{\left(1000 \dfrac{g}{L}\right)\left(4.184 \dfrac{J}{g°C}\right)}$$
$$= 25.1°C \quad (25°C)$$

The temperature after adding NaOH is

$$39°C + 25°C = 64°C$$

The answer is (C).

42. The total solid waste generated by the community is

$$(15\,000 \text{ people})\left(1.6 \dfrac{kg}{person \cdot d}\right) = 24\,000 \text{ kg/d}$$

The answer is (D).

43. Assume a 100 kg sample.

waste component	discarded mass (kg)	dry mass (kg)
food	13	3.9
glass	6	5.9
plastic	4	3.9
paper	37	35
cardboard	10	9.5
textiles	1	0.9
ferrous metal	8	7.8
non-ferrous metal	2	2.0
wood	4	3.2
yard clippings	15	6.0
	100	78.1

$$\text{discarded mass, kg} = (100 \text{ kg})\left(\dfrac{\% \text{ mass}}{100}\right)$$

$$\text{dry mass, kg} = \dfrac{(\text{discarded mass, kg}) \times (100 - \% \text{ moisture})}{100}$$

$$(100 \text{ kg} - 78.1 \text{ kg})\left(\dfrac{100\%}{100 \text{ kg}}\right) = 21.9\% \quad (22\%)$$

The answer is (B).

44. Assume a 100 kg sample.

waste component	discarded mass (kg)	discarded volume (m³)
food	13	0.04
glass	6	0.031
plastic	4	0.062
paper	37	0.44
cardboard	10	0.20
textiles	1	0.015
ferrous metal	8	0.025
non-ferrous metal	2	0.013
wood	4	0.017
yard clippings	15	0.14
	100	0.983

$$\text{discarded mass, kg} = (100 \text{ kg})\left(\frac{\% \text{ mass}}{100}\right)$$

$$\text{discarded volume, m}^3 = \frac{\text{discarded mass, kg}}{\text{discarded density, kg/m}^3}$$

$$\frac{\text{bulk discarded}}{\text{density, kg/m}^3} = \frac{100 \text{ kg}}{0.983 \text{ m}^3}$$

$$= 101.7 \text{ kg/m}^3 \quad (100 \text{ kg/m}^3)$$

The answer is (B).

45. Assume a 100 kg sample.

waste component	discarded mass (kg)	unit discarded energy (10^6 kJ/100 kg)	ash mass (kg/100 kg)
food	13	0.060	0.65
glass	6	0.00090	5.9
plastic	4	0.13	0.4
paper	37	0.62	2.2
cardboard	10	0.16	0.50
textiles	1	0.017	0.025
ferrous metal	8	0.0056	7.8
non-ferrous metal	2	0.0014	1.9
wood	4	0.074	0.060
yard clippings	15	0.098	0.68
	100	1.17	20

$$\text{discarded mass, kg} = (100 \text{ kg})\left(\frac{\% \text{ mass}}{100}\right)$$

$$\begin{aligned}\text{unit discarded energy,} \\ 10^6 \text{ kJ/100 kg}\end{aligned} = (\text{discarded mass, kg})$$

$$\times \left(\begin{array}{c}\text{component discarded} \\ \text{energy, kJ/kg}\end{array}\right)$$

$$\text{ash mass, kg/100 kg} = (\text{discarded mass, kg})$$

$$\times \left(\frac{\% \text{ ash}}{100}\right)$$

$$\left(\frac{1.17 \times 10^6 \text{ kJ}}{100 \text{ kg discarded mass}}\right)\left(\frac{100 \text{ kg discarded mass}}{80 \text{ kg discarded mass without ash}}\right)$$

$$\times \left(\frac{(10)(100 \text{ kg})}{1000 \text{ kg}}\right)$$

$$= 14.6 \times 10^6 \text{ kJ/1000 kg} \quad (1.5 \times 10^7 \text{ kJ/1000 kg})$$

The answer is (A).

46.

$$C_aH_bO_cN_d + \frac{(4a - b - 2c + 3d)}{4}H_2O$$
$$\rightarrow \frac{(4a + b - 2c - 3d)}{8}CH_4$$
$$+ \frac{(4a - b + 2c + 3d)}{8}CO_2 + dNH_3$$

$$a = 81$$
$$b = 179$$
$$c = 68$$
$$d = 1$$

$$C_{81}H_{179}O_{68}N + 3H_2O \rightarrow 45.5CH_4 + 35.5CO_2 + NH_3$$

1 mol of $C_{81}H_{179}O_{68}N$ produces 45.5 mol of CH_4.

$$CH_4 \text{ MW} = 12 \frac{\text{g}}{\text{mol}} + (4)\left(1 \frac{\text{g}}{\text{mol}}\right) = 16 \text{ g/mol}$$

$$\frac{(45.5 \text{ mol CH}_4)\left(16 \frac{\text{g}}{\text{mol}}\right)}{(1 \text{ mol waste})\begin{pmatrix}(81)\left(12 \frac{\text{g}}{\text{mol}}\right) \\ + (179)\left(1 \frac{\text{g}}{\text{mol}}\right) \\ + (68)\left(16 \frac{\text{g}}{\text{mol}}\right) \\ + (1)\left(14 \frac{\text{g}}{\text{mol}}\right)\end{pmatrix}}$$

$$= 0.32 \text{ g CH}_4/\text{g waste}$$
$$(0.32 \text{ kg CH}_4/\text{kg waste})$$

$$\left(0.32 \frac{\text{kg}}{\text{kg}}\right)(13\,000 \text{ people})\left(1.2 \frac{\text{kg}}{\text{person} \cdot \text{d}}\right)$$
$$= 4992 \text{ kg CH}_4/\text{d} \quad (5000 \text{ kg CH}_4/\text{d})$$

The answer is (B).

47. Over 90% of the gas volume produced from anaerobic decomposition of solid waste in landfills consists of methane (CH_4) and carbon dioxide (CO_2).

The answer is (B).

48. The compacted waste without cover is

$$\frac{(13\,000 \text{ people})\left(1.2 \dfrac{\text{kg}}{\text{person} \cdot \text{d}}\right)}{1100 \dfrac{\text{kg}}{\text{m}^3}} = 14.2 \text{ m}^3/\text{d}$$

Using 1 m³ of cover for every 5 m³ of waste, the compacted waste with cover is

$$\frac{14.2 \dfrac{\text{m}^3}{\text{d}}}{5} + 14.2 \dfrac{\text{m}^3}{\text{d}} = 17 \text{ m}^3/\text{d}$$

Assume 365 d/yr of operation.

$$\left(17 \dfrac{\text{m}^3}{\text{d}}\right)\left(365 \dfrac{\text{d}}{\text{yr}}\right)(25 \text{ yr}) = 160 \times 10^3 \text{ m}^3$$

The answer is (A).

49. The material to be burned has small particle size and likely low ash. Therefore, a fluid bed incinerator would be best suited for the waste.

The answer is (C).

50. Controlled air incinerators are relatively low-technology devices that have difficulty meeting air emission limits.

The answer is (A).

51. The minimum required bed incinerator area is

$$\frac{(\text{HV})(\text{fueling rate})}{\text{heat release rate}} = \frac{\left(500 \dfrac{\text{kg}}{\text{h}}\right)\left(57\,000 \dfrac{\text{kJ}}{\text{kg}}\right)}{2.3 \times 10^6 \dfrac{\text{kJ}}{\text{m}^2 \cdot \text{h}}}$$

$$= 12.4 \text{ m}^2 \quad (12 \text{ m}^2)$$

The answer is (C).

52. The four primary factors used to control incineration efficiency are burn temperature, oxygen level, turbulence, and residence time.

The answer is (A).

53.
$$\text{CrO}_3 \text{ MW} = 52 \dfrac{\text{g}}{\text{mol}} + (3)\left(16 \dfrac{\text{g}}{\text{mol}}\right)$$
$$= 100 \text{ g/mol} \quad (100 \text{ mg/mmol})$$

$$\frac{534 \dfrac{\text{mg}}{\text{L}}}{100 \dfrac{\text{mg}}{\text{mmol}}} = 5.34 \dfrac{\text{mmol}}{\text{L}}$$

$$\text{NaHSO}_3 \text{ MW} = 23 \dfrac{\text{g}}{\text{mol}} + 1 \dfrac{\text{g}}{\text{mol}} + 32 \dfrac{\text{g}}{\text{mol}}$$
$$+ (3)\left(16 \dfrac{\text{g}}{\text{mol}}\right)$$
$$= 104 \text{ g/mol} \quad (104 \text{ mg/mmol})$$

From the reduction reaction, (4/4)(5.34 mmol/L) of CrO_3 will react with (6/4)(5.34 mmol/L) of NaHSO_3.

$$\left(\dfrac{6}{4}\right)\left(5.34 \dfrac{\text{mmol}}{\text{L}}\right) = 8.01 \text{ mmol/L}$$

The facility operates 24 h/d for 365 d/yr.

$$\left(104 \dfrac{\text{mg}}{\text{mmol}}\right)\left(8.01 \dfrac{\text{mmol}}{\text{L}}\right)\left(1.8 \dfrac{\text{m}^3}{\text{min}}\right)$$
$$\times \left(10^{-3} \dfrac{\text{kg} \cdot \text{L}}{\text{mg} \cdot \text{m}^3}\right)\left(1440 \dfrac{\text{min}}{\text{d}}\right)\left(365 \dfrac{\text{d}}{\text{yr}}\right)$$
$$= 7.88 \times 10^5 \text{ kg/yr} \quad [\text{for 100\% pure NaHSO}_3]$$

$$\frac{\left(7.88 \times 10^5 \dfrac{\text{kg}}{\text{yr}}\right)\left(\dfrac{\$120}{1000 \text{ kg}}\right)}{0.92}$$
$$= \$102,870/\text{yr} \quad (\$100,000/\text{yr})$$

The answer is (C).

54.
$$\text{CrO}_3 \text{ MW} = 52 \dfrac{\text{g}}{\text{mol}} + (3)\left(16 \dfrac{\text{g}}{\text{mol}}\right)$$
$$= 100 \text{ g/mol} \quad (100 \text{ mg/mmol})$$

$$\frac{534 \dfrac{\text{mg}}{\text{L}}}{100 \dfrac{\text{mg}}{\text{mmol}}} = 5.34 \text{ mmol/L}$$

$$\text{Cr}_2(\text{SO}_4)_3 \text{ MW} = (2)\left(52 \dfrac{\text{g}}{\text{mol}}\right) + (3)\left(\begin{array}{l} 32 \dfrac{\text{g}}{\text{mol}} + (4) \\ \times \left(16 \dfrac{\text{g}}{\text{mol}}\right) \end{array}\right)$$
$$= 392 \text{ g/mol} \quad (392 \text{ mg/mmol})$$

From the reduction reaction, (4/4)(5.34 mmol/L) of CrO_3 will react to produce (2/4)(5.34 mmol/L) of $\text{Cr}_2(\text{SO}_4)_3$.

$$\left(\dfrac{2}{4}\right)\left(5.34 \dfrac{\text{mmol}}{\text{L}}\right) = 2.67 \text{ mmol/L}$$

The facility operates 24 h/d for 365 d/yr.

$$\left(392 \dfrac{\text{mg}}{\text{mmol}}\right)\left(2.67 \dfrac{\text{mmol}}{\text{L}}\right)\left(1.8 \dfrac{\text{m}^3}{\text{min}}\right)$$
$$\times \left(10^{-3} \dfrac{\text{kg} \cdot \text{L}}{\text{mg} \cdot \text{m}^3}\right)\left(1440 \dfrac{\text{min}}{\text{d}}\right)\left(365 \dfrac{\text{d}}{\text{yr}}\right)$$
$$= 9.9 \times 10^5 \text{ kg/yr} \quad [\text{dry mass}]$$

The answer is (C).

55. Because the combined capacity of all the tanks is much greater than 1320 gal and all tanks contain oil or oil products, all tanks are subject to inclusion in an SPCC Plan.

The answer is (A).

56. The containment volume based on the largest tank is

$$(1{,}000{,}000 \text{ gal})(1.1)\left(0.134 \ \frac{\text{ft}^3}{\text{gal}}\right) = 147{,}400 \text{ ft}^3$$

The containment volume based on storm water is

$$(1.85 \text{ ac})(14 \text{ in})\left(43{,}560 \ \frac{\text{ft}^2}{\text{ac}}\right)\left(\frac{1 \text{ ft}}{12 \text{ in}}\right) = 94{,}017 \text{ ft}^3$$

The berm height without freeboard is

$$\frac{147{,}400 \text{ ft}^3 + 94{,}017 \text{ ft}^3}{(1.85 \text{ ac})\left(43{,}560 \ \frac{\text{ft}^2}{\text{ac}}\right)} = 3.0 \text{ ft}$$

The berm height with freeboard is

$$3.0 \text{ ft} + 1.5 \text{ ft} = 4.5 \text{ ft}$$

The answer is (C).

57. The three major sources of air pollution in the United States are transportation, fuel combustion not associated with vehicles (such as coal-fired power plants), and industrial processes not associated with burning fuel.

The answer is (D).

58. Applying Dalton's law of partial pressures, the partial pressure exerted by argon is

$$\left(\frac{0.93\%}{100\%}\right)(0.98 \text{ atm}) = 0.0091 \text{ atm}$$

The answer is (A).

59.

component	fractional composition	molecular weight (g/mol)	fractional molecular weight (g/mol)
nitrogen, N_2	0.7809	28	21.865
oxygen, O_2	0.2094	32	6.7008
argon, Ar	0.0093	40	0.3720
carbon dioxide, CO_2	0.000315	44	0.01386
			28.952

The apparent molecular weight is 28.952 g/mol (29 g/mol).

The answer is (A).

60.

$$O_3 \text{ MW} = (3)\left(16 \ \frac{\text{g}}{\text{mol}}\right) = 48 \text{ g/mol}$$

From the universal gas law, the molar volume of air at 1 atm and 24°C is

$$\frac{\left(8.2 \times 10^{-5} \ \frac{\text{atm} \cdot \text{m}^3}{\text{mol} \cdot \text{K}}\right)(24°\text{C} + 273°)}{1 \text{ atm}} = 0.0244 \text{ m}^3/\text{mol}$$

The concentration is

$$\frac{\left(0.0244 \ \frac{\text{m}^3}{\text{mol}}\right)\left(213 \ \frac{\mu\text{g}}{\text{m}^3}\right)}{48 \ \frac{\text{g}}{\text{mol}}} = 0.1083 \ \frac{\mu\text{g}}{\text{g}}$$

$$= 0.1083 \text{ ppm}$$

$$(0.11 \text{ ppm})$$

The answer is (C).

61. Air quality monitoring occurs by applying three different categories of measurements: emissions, ambient air, and meteorological. Emissions measurements apply to both stationary and mobile sources and occur by sampling or monitoring at the point where the emission leaves the source. Ambient air is measured to provide background to compare with emissions monitoring results and to assess pollutant levels. Meteorological measurements include wind speed and direction, air temperature profiles, and other parameters necessary to evaluate pollutant dispersion and fate.

The answer is (A).

62. For each pollutant included in the NAAQS, primary and secondary standards are defined. Primary standards are intended to protect human health, and secondary standards are intended to protect public welfare.

Air pollutants may be classified as primary and secondary. Primary pollutants are those that exist in the air in the same form in which they were emitted. Secondary pollutants are those that are formed in the air from other emitted compounds. Primary and secondary pollutants are not the same as, and should not be confused with, the primary and secondary NAAQS.

The answer is (B).

63. The relevant basic equations for electrostatic precipitator design include the following.

$$w = \sqrt{\frac{4g\rho_p d_d}{3C_D \rho_a}}$$

$$g = \frac{6\epsilon_o E^2}{d_d \rho_p}$$

$$C_D = \frac{24}{\text{Re}}$$

$$\text{Re} = \frac{d_d w \rho_a}{\mu_g}$$

Combining equations and canceling common terms yields

$$w = \frac{8\epsilon_o E^2 d_d}{24\,\mu_g}$$

For a gas stream temperature of 176°C, absolute viscosity is

$$\mu_g = 2.5 \times 10^{-5} \text{ kg/m·s}$$

The permittivity constant is

$$\epsilon_o = 8.85 \times 10^{-12} \text{ C}^2/\text{N·m}^2$$

The particle drift velocity is

$$w = \frac{(8)\left(8.85 \times 10^{-12}\,\frac{\text{C}^2}{\text{N·m}^2}\right)\left(700\,000\,\frac{\text{N}}{\text{C}}\right)^2 (4\,\mu\text{m})}{\left(10^6\,\frac{\mu\text{m}}{\text{m}}\right)\left(2.5 \times 10^{-5}\,\frac{\text{kg}}{\text{m·s}}\right)\left(\frac{\text{N·s}^2}{\text{kg·m}}\right)(24)}$$

$$= 0.23 \text{ m/s}$$

The answer is (A).

64. The plate area is

$$A = \frac{-\ln\left(\frac{1-E\%}{100\%}\right) Q_g}{w}$$

$$= \frac{-\ln\left(\frac{1-80\%}{100\%}\right)\left(14\,\frac{\text{m}^3}{\text{s}}\right)}{1.0\,\frac{\text{m}}{\text{s}}}$$

$$= 22.53 \text{ m}^2 \quad (23 \text{ m}^2)$$

The answer is (C).

65. The mass removed is

$$\frac{C_p Q_g E\%}{100\%} = \frac{\left(20\,\frac{\text{g}}{\text{m}^3}\right)\left(14\,\frac{\text{m}^3}{\text{s}}\right)(80\%)}{(100\%)\left(\frac{1\,\text{d}}{86\,400\,\text{s}}\right)\left(1000\,\frac{\text{g}}{\text{kg}}\right)}$$

$$= 19\,354 \text{ kg/d} \quad (19\,000 \text{ kg/d})$$

The answer is (D).

66. For air at 30°C, the absolute viscosity is

$$\mu_g = 1.9 \times 10^{-5} \text{ kg/m·s}$$

Assuming MW for air of 29 g/mol, the air density at 1.3 atm is

$$\rho_g = \frac{(\text{MW})(p)}{TR^*}$$

$$= \frac{\left(29\,\frac{\text{g}}{\text{mol}}\right)(1.3\,\text{atm})\left(\frac{1\,\text{m}^3}{10^6\,\text{cm}^3}\right)}{(30°\text{C} + 273°)\left(8.2 \times 10^{-5}\,\frac{\text{atm·m}^3}{\text{mol·K}}\right)}$$

$$= 0.001\,52 \text{ g/cm}^3$$

Applying Stoke's law, the particle settling velocity is

$$v_s = \frac{d_p^2(\rho_p - \rho_g)g}{18\,\mu_g}$$

$$= \frac{(2.5\,\mu\text{m})^2\left(10^{-6}\,\frac{\text{m}}{\mu\text{m}}\right)^2}{(18)\left(1.9 \times 10^{-5}\,\frac{\text{kg}}{\text{m·s}}\right)\left(10^3\,\frac{\text{g}}{\text{kg}}\right)}$$

$$\times \left(0.58\,\frac{\text{g}}{\text{cm}^3} - 0.00152\,\frac{\text{g}}{\text{cm}^3}\right)$$

$$\times \left(100\,\frac{\text{cm}}{\text{m}}\right)^4 \left(9.81\,\frac{\text{m}}{\text{s}^2}\right)$$

$$= 0.0104 \text{ cm/s} \quad (0.010 \text{ cm/s})$$

The answer is (B).

67. For air at 30°C, the absolute viscosity is

$$\mu_g = 1.9 \times 10^{-5} \text{ kg/m·s}$$

Assuming MW for air of 29 g/mol, air density at 1.3 atm is

$$\rho_g = \frac{(\text{MW})(p)}{TR^*}$$

$$= \frac{\left(29\,\frac{\text{g}}{\text{mol}}\right)(1.3\,\text{atm})\left(\frac{1\,\text{m}^3}{10^6\,\text{cm}^3}\right)}{(30°\text{C} + 273°)\left(8.2 \times 10^{-5}\,\frac{\text{atm·m}^3}{\text{mol·K}}\right)}$$

$$= 0.001\,52 \text{ g/cm}^3$$

The Reynolds number is

$$\text{Re} = \frac{v_s d_p \rho_g}{\mu_g}$$

$$= \frac{\left(0.10\,\frac{\text{m}}{\text{s}}\right)(2.5\,\mu\text{m})\left(10^{-6}\,\frac{\text{m}}{\mu\text{m}}\right)}{\left(1.9 \times 10^{-5}\,\frac{\text{kg}}{\text{m·s}}\right)\left(10^3\,\frac{\text{g}}{\text{kg}}\right)}$$

$$\times \left(0.001\,52\,\frac{\text{g}}{\text{cm}^3}\right)\left(100\,\frac{\text{cm}}{\text{m}}\right)^3$$

$$= 0.020$$

The answer is (C).

68. Assume the gas stream to be dry. The partial pressure of the solvent vapor in the gas stream is

$$180 \text{ ppm} = \frac{180 \text{ g}}{10^6 \text{ g}}$$

$$p = \left(\frac{180 \text{ g}}{10^6 \text{ g}}\right)(1\,\text{atm})\left(101.3\,\frac{\text{kPa}}{\text{atm}}\right)$$

$$\times \left(10^3\,\frac{\text{Pa}}{\text{kPa}}\right)$$

$$= 18.23 \text{ Pa}$$

The GAC adsorption capacity is determined using the Freundlich isotherm equation.

$$\frac{X}{M} = K_f p^{1/n} = (0.126)(18.23 \text{ Pa})^{0.176}$$
$$= 0.210 \text{ g solvent vapor/g GAC}$$

The molar gas volume at 25°C is

$$V = \frac{uR^*T}{p}$$

$$= \frac{(1 \text{ mol})\left(8.2 \times 10^{-5} \, \frac{\text{atm·m}^3}{\text{mol·K}}\right)(25°\text{C} + 273°)}{(1 \text{ atm})\left(10^{-3} \, \frac{\text{m}^3}{\text{L}}\right)}$$

$$= 24.44 \text{ L/mol}$$

The solvent vapor concentration is

$$\frac{\left(\frac{180 \text{ g}}{10^6 \text{ g}}\right)\left(78 \, \frac{\text{g}}{\text{mol}}\right)\left(10^3 \, \frac{\text{L}}{\text{m}^3}\right)}{\left(24.44 \, \frac{\text{L}}{\text{mol}}\right)} = 0.574 \text{ g/m}^3$$

The solvent vapor mass removed from the gas stream is

$$\left(4.2 \, \frac{\text{m}^3}{\text{s}}\right)\left(0.574 \, \frac{\text{g}}{\text{m}^3}\right)\left(86\,400 \, \frac{\text{s}}{\text{d}}\right) = 2.08 \times 10^5 \text{ g/d}$$

The GAC use rate is

$$\frac{\left(2.08 \times 10^5 \, \frac{\text{g}}{\text{d}}\right)\left(\frac{1 \text{ kg}}{10^3 \text{ g}}\right)}{0.210 \, \frac{\text{g}}{\text{g GAC}}} = 990 \text{ kg GAC/d}$$
$$(990 \text{ kg/d})$$

The answer is (B).

69. For a regeneration cycle of 1/d, assume the bed mass is 1000 kg.

The GAC bed area is

$$\frac{(1000 \text{ kg})\left(10^3 \, \frac{\text{g}}{\text{kg}}\right)\left(\frac{1 \text{ m}}{100 \text{ cm}}\right)^2}{\left(0.40 \, \frac{\text{g}}{\text{cm}^3}\right)(60 \text{ cm})} = 4.2 \text{ m}^2$$

The answer is (B).

70. The gas flow velocity through the bed per 10 m² of area is

$$v = \frac{Q}{A}$$

$$v_g = \frac{4.2 \, \frac{\text{m}^3}{\text{s}}}{10 \text{ m}^2}$$
$$= 0.42 \text{ m/s}$$

The friction factor for a Reynolds number of 1000 is

$$f = \frac{150(1-\alpha)}{\text{Re}} + 1.75$$
$$= \frac{(150)(1-0.38)}{1.0 \times 10^3} + 1.75$$
$$= 1.843$$

The geometric mean particle diameter is

no. 6 sieve opening = 3.35 mm
no. 12 sieve opening = 1.70 mm

$$d_g = \frac{\sqrt{(3.35 \text{ mm})(1.70 \text{ mm})}}{1000 \, \frac{\text{mm}}{\text{m}}}$$

$$= 0.00239 \text{ m}$$

Because 100% of the media falls between no. 6 mesh and no. 12 sieves mesh, the mass fraction of particles between adjacent sieves is

$$p = 1$$

The head loss through the bed is

$$h = \frac{L(1-\alpha)v_g^2 f p}{d_g \alpha^3 g}$$

$$= \frac{(60 \text{ cm})(1-0.38)\left(0.42 \, \frac{\text{m}}{\text{s}}\right)^2 (1.843)(1)}{(0.00239 \text{ m})(0.38)^3 \left(9.81 \, \frac{\text{m}}{\text{s}^2}\right)\left(100 \, \frac{\text{cm}}{\text{m}}\right)}$$

$$= 94 \text{ m}$$

The answer is (C).

71. The average temperature change through the condenser is

$$\Delta T_m = \frac{(T_{1a} - T_{2l}) - (T_{2a} - T_{1l})}{\ln\left(\frac{T_{1a} - T_{2l}}{T_{2a} - T_{1l}}\right)}$$

$$= \frac{(198°\text{F} - 130°\text{F}) - (80°\text{F} - 40°\text{F})}{\ln\left(\frac{198°\text{F} - 130°\text{F}}{80°\text{F} - 40°\text{F}}\right)}$$

$$= 53°\text{F}$$

The answer is (B).

72. Carbon monoxide will cause death in humans within a few minutes at exposures exceeding 5000 ppm and impair vision, motor function, and other physical and mental abilities at much lower concentrations. These effects occur through formation of carboxyhemoglobin as a reaction product between carbon monoxide and hemoglobin. The carboxyhemoglobin formation effectively acts to deprive the body of oxygen. Heightened nervousness and mental agitation are typically not associated with carbon monoxide exposure.

The answer is (B).

73. Carbon monoxide is an odorless and colorless gas that is formed by the incomplete combustion of fossil fuels and other organic compounds. In the atmosphere, carbon monoxide may react with hydroxyl radicals to form carbon dioxide or it may be removed by soil organisms. Both of these mechanisms are thought to explain why atmospheric carbon monoxide levels have remained essentially unchanged for decades even though fossil fuel consumption has increased.

The answer is (D).

74. The pollutant standard index is determined using published PSI health impact index and breakpoint values.

pollutant	concentration	upper index	lower index	upper breakpoint	lower breakpoint
O_3	0.21 ppm	300	200	0.40 ppm	0.20 ppm
CO	11 ppm	200	100	15 ppm	9 ppm
PM_{10}	240 $\mu g/m^3$	200	100	350 $\mu g/m^3$	150 $\mu g/m^3$
SO_2	0.33 ppm	300	200	0.60 ppm	0.30 ppm
NO_2	0.8 ppm	300	200	1.2 ppm	0.6 ppm

The subindex for each pollutant is found using linear interpolation.

$$\text{subindex } O^3 = 200 + (300 - 200)\left(\frac{0.21 - 0.20}{0.40 - 0.20}\right)$$
$$= 205$$

$$\text{subindex CO} = 100 + (200 - 100)\left(\frac{11 - 9}{15 - 9}\right)$$
$$= 133$$

$$\text{subindex PM}_{10} = 100 + (200 - 100)\left(\frac{240 - 150}{350 - 150}\right)$$
$$= 145$$

$$\text{subindex SO}_2 = 200 + (300 - 200)\left(\frac{0.33 - 0.30}{0.60 - 0.30}\right)$$
$$= 210$$

$$\text{subindex NO}_2 = 200 + (300 - 200)\left(\frac{0.8 - 0.6}{1.2 - 0.6}\right)$$
$$= 233$$

The highest subindex is calculated for NO_2.
$$\text{PSI} = 233$$

The answer is (D).

75. For a pollutant standard index of 230, the air quality descriptor is very unhealthful, as the following table shows.

PSI value	descriptor
0–49	good
50–99	moderate
100–199	unhealthful
200–299	very unhealthful
300–399	hazardous (at-risk persons should stay indoors)
> 400	hazardous (all persons should remain indoors)

The answer is (C).

76. Wind is measured from the direction of origin, and prevailing wind is the wind with the highest percentage of occurrence. The prevailing wind originates from the south.

The answer is (B).

77. Wind originating from the south-southwest exceeds 10 mph approximately 8% of the time.

The answer is (C).

78. For carcinogens, it is assumed that any dose will result in a risk. For noncarcinogens, a threshold dose is assumed, below which no risk occurs.

The answer is (C).

79. The potency factor (PF) of a toxic chemical is the slope of the linear portion of the dose-response curve. The potency factor is sometimes also referred to as the slope factor.

The answer is (B).

80. Group A compounds are those for which data from animal tests or from human epidemiological studies are sufficient to define a correlation between exposure to the chemical and a risk of contracting cancer.

The answer is (D).

81. The reference dose (RfD) defines the dose of a noncarcinogen that will produce an unacceptable risk to a population upon exposure.

The answer is (D).

82. Bioconcentration of toxic contaminants in fish is evaluated using the bioconcentration factor (BCF).

The answer is (C).

83.
$$E_c = \frac{C_1 t_1 + C_2 t_2 + C_3 t_3}{8 \text{ h}}$$

E_c	cumulative exposure	ppm
C	concentration 1, 2, 3 of chemical	ppm
t	time of exposure 1, 2, 3 to chemical	h

$$E_c = \frac{(12 \text{ ppm})(3 \text{ h}) + (4 \text{ ppm})(5 \text{ h})}{8 \text{ h}} = 7 \text{ ppm}$$

The answer is (C).

84.
$$E_c = \frac{C_1 t_1 + C_2 t_2 + C_3 t_3}{8 \text{ h}}$$

$$E_{\text{acetone}} = \frac{(1080 \text{ ppm})(1 \text{ h}) + (630 \text{ ppm}) \times (2 \text{ h}) + (470 \text{ ppm})(5 \text{ h})}{8 \text{ h}}$$
$$= 586 \text{ ppm}$$

$$E_{\text{sec-butanol}} = \frac{(180 \text{ ppm})(1 \text{ h}) + (85 \text{ ppm}) \times (2 \text{ h}) + (15 \text{ ppm})(5 \text{ h})}{8 \text{ h}}$$
$$= 53 \text{ ppm}$$

$$E_m = \frac{E_{c1}}{L_1} + \frac{E_{c2}}{L_2}$$

E_m	cumulative exposure to mixture	–
L	8 h TWA PEL for chemical 1 and 2	ppm

$$E_m = \frac{586 \text{ ppm}}{1000 \text{ ppm}} + \frac{53 \text{ ppm}}{150 \text{ ppm}} = 0.94$$

The answer is (B).

85. $E_m = 1.6$ is greater than 1.0. Therefore, exposure to the mixture of acetone, benzene, and sec-butanol is not acceptable.

The answer is (D).

86.
$$E_c = \frac{(42 \text{ ppm})(8 \text{ min})\left(\frac{1 \text{ h}}{60 \text{ min}}\right) + (12 \text{ ppm}) \times \left(3 \text{ h} - (8 \text{ min})\left(\frac{1 \text{ h}}{60 \text{ min}}\right)\right) + (4 \text{ ppm})(5 \text{ h})}{8 \text{ h}}$$
$$= 7.5 \text{ ppm}$$

The 8 h TWA PEL for benzene is 10 ppm. Therefore, the exposure is acceptable because 7.5 ppm is less than 10 ppm.

The answer is (C).

87. All choices are statutes relevant to nuclear energy and radiation, but the Radiation Control for Health and Safety Act of 1968 specifically applies to radiation exposure from X rays and other medical uses associated with radiation.

The answer is (D).

88. The quality factor, Q_F, for alpha particles is 20. The dose equivalent is

$$h = \text{absorbed} = (0.13 \text{ rad})(20) = 2.6 \text{ rem}$$

The answer is (D).

89. The intensity at 10 m is

$$I_{10} = \frac{I_2 d_2^2}{d_{10}^2}$$
$$= \frac{\left(0.016 \dfrac{\text{R}}{\text{min}}\right)(2 \text{ m})^2 \left(1000 \dfrac{\text{mR}}{\text{R}}\right)}{(10 \text{ m})^2}$$
$$= 0.64 \text{ mR/min}$$

The answer is (B).

90. Chemical A has a low MCL and a low Henry's constant compared to chemicals B, C, and D. The lower MCL requires a higher removal efficiency. The lower Henry's constant makes the removal efficiency more difficult to achieve. Chemical A should be the target contaminant.

The answer is (A).

91. The mass of chemical A released is

$$m_A = (1500 \text{ gal})\left(3.785 \dfrac{\text{L}}{\text{gal}}\right)\left(0.72 \dfrac{\text{kg gasoline}}{\text{kg water}}\right)$$
$$\times \left(1 \dfrac{\text{kg water}}{\text{L}}\right)\left(0.018 \dfrac{\text{g}}{\text{g}}\right)$$
$$= 73.6 \text{ kg} \quad (74 \text{ kg})$$

The answer is (B).

92. The vapor concentration of chemical C is

$$C_v = \frac{x \rho_g (\text{MW})}{R^* T}$$

MW	molecular weight	g/mol
ρ_g	vapor pressure	atm
R^*	universal gas constant	cm$^3 \cdot$atm/mol$\cdot$K
T	ambient soil temperature	K
x	mole fraction	–

$$MW = 106 \text{ g/mol}$$
$$R^* = 82.1 \text{ cm}^3 \cdot \text{atm/mol} \cdot \text{K}$$
$$T = 281 \text{K}$$
$$x = 0.018$$
$$\rho_g = 0.0092 \text{ atm}$$

$$C_v = \frac{(0.018)(0.0092 \text{ atm})\left(106 \frac{\text{g}}{\text{mol}}\right)\left(10^{-3} \frac{\text{kg}}{\text{g}}\right)}{\left(82.1 \frac{\text{cm}^3 \cdot \text{atm}}{\text{mol} \cdot \text{K}}\right)(281 \text{K})\left(\frac{1 \text{ m}^3}{10^6 \text{ cm}^3}\right)}$$
$$= 0.00076 \text{ kg/m}^3$$

The emission rate is the vent flow rate multiplied by C_v.

$$\left(20 \frac{\text{ft}^3}{\text{min}}\right)\left(0.00076 \frac{\text{kg}}{\text{m}^3}\right)\left(1440 \frac{\text{min}}{\text{d}}\right)\left(\frac{1 \text{ m}^3}{35.3 \text{ ft}^3}\right)$$
$$= 0.62 \text{ kg/d}$$

The answer is (D).

93. The mass of chemical B extracted is

$$m_B = (1500 \text{ gal})\left(3.785 \frac{\text{L}}{\text{gal}}\right)\left(0.72 \frac{\text{kg gasoline}}{\text{kg water}}\right)$$
$$\times \left(1 \frac{\text{kg water}}{\text{L}}\right)\left(0.082 \frac{\text{g}}{\text{g}}\right)(0.90)$$
$$= 301.7 \text{ kg} \quad (300 \text{ kg})$$

The answer is (D).

94. The mass of chemical D emitted is

$$\frac{(1500 \text{ gal})\left(3.785 \frac{\text{L}}{\text{gal}}\right)\left(0.72 \frac{\text{kg gasoline}}{\text{kg water}}\right)}{(8 \text{ mo})\left(30 \frac{\text{d}}{\text{mo}}\right)\left(20 \frac{\text{ft}^3}{\text{min}}\right)\left(1440 \frac{\text{min}}{\text{d}}\right)}$$
$$\times \left(\frac{1 \text{ m}^3}{35.3 \text{ ft}^3}\right)\left(\frac{1 \text{ kg}}{10^3 \text{ g}}\right)$$
$$= 0.027 \text{ g/m}^3$$

The answer is (A).

95. Radon gas is emitted from water during normal household uses, but only when the water supply is untreated groundwater. Treated water poses little risk of radon exposure.

The answer is (A).

96. The two treatment options for radon removal from groundwater are aeration and granular activated carbon adsorption. Aeration is a best available technology (BAT).

The answer is (C).

97. The major indoor air pollutants include a variety of substances, but of these, radon and environmental tobacco smoke (ETS) are the most significant because of their relative toxicity, frequency of occurrence, and potential to exist at elevated concentrations.

The answer is (D).

98. Nonoccupational exposure to asbestos has been linked to lung cancer and mesothelioma. Asbestosis is typically associated with occupational exposure. Black lung and cystic fibrosis have not been linked to asbestos exposure.

The answer is (D).

99.

T temperature   5°C
t contact time   30 min

The free chlorine concentration in mg/L is

$$C = \frac{0.9847 C^{0.1758}(\text{pH})^{2.7519} T^{-0.1467}}{t}$$
$$= \frac{0.9847 C^{0.1758}(7.7)^{2.7519}(5°\text{C})^{-0.1467}}{30 \text{ min}}$$
$$= 10.8 \text{ mg/L} \quad (11 \text{ mg/L})$$

The answer is (D).

100. The optimum pH for chlorine disinfection is between 6.5 and 7.5. The minimum dose will occur at the lowest pH within the range. Use pH = 6.5.

T temperature   5°C
t contact time   30 min

The free chlorine concentration in mg/L is

$$C = \frac{0.9847 C^{0.1758}(\text{pH})^{2.7519} T^{-0.1467}}{t}$$
$$= \frac{0.9847 C^{0.1758}(6.5)^{2.7519}(5°\text{C})^{-0.1467}}{30 \text{ min}}$$
$$= 6.16 \text{ mg/L} \quad (6.2 \text{ mg/L})$$

The answer is (A).

Instructions

Name: _____
 Last   First   Middle Initial

Do not enter solutions in the test booklet. Complete solutions must be entered on the answer sheet provided by your proctor.

This is an open-book examination. You may use textbooks, handbooks, and other bound references, along with a battery-operated, silent, nonprinting calculator. Unbound reference materials and notes, scratchpaper, and writing tablets are not permitted. You may not consult with or otherwise share any materials or information with others taking the exam.

You must work all 50 multiple-choice problems in the four-hour period allocated for the morning session. Each of the 50 problems is worth 1 point. No partial credit will be awarded. Your score will be based entirely on the responses marked on the answer sheet. You may use blank spaces in the exam booklet for scratch work. However, no credit will be awarded for work shown in margins or on other pages of the exam booklet. Mark only one answer to each problem.

Principles and Practice of Engineering Examination

MORNING SESSION
Sample Examination 2

101. Ⓐ Ⓑ Ⓒ Ⓓ	126. Ⓐ Ⓑ Ⓒ Ⓓ
102. Ⓐ Ⓑ Ⓒ Ⓓ	127. Ⓐ Ⓑ Ⓒ Ⓓ
103. Ⓐ Ⓑ Ⓒ Ⓓ	128. Ⓐ Ⓑ Ⓒ Ⓓ
104. Ⓐ Ⓑ Ⓒ Ⓓ	129. Ⓐ Ⓑ Ⓒ Ⓓ
105. Ⓐ Ⓑ Ⓒ Ⓓ	130. Ⓐ Ⓑ Ⓒ Ⓓ
106. Ⓐ Ⓑ Ⓒ Ⓓ	131. Ⓐ Ⓑ Ⓒ Ⓓ
107. Ⓐ Ⓑ Ⓒ Ⓓ	132. Ⓐ Ⓑ Ⓒ Ⓓ
108. Ⓐ Ⓑ Ⓒ Ⓓ	133. Ⓐ Ⓑ Ⓒ Ⓓ
109. Ⓐ Ⓑ Ⓒ Ⓓ	134. Ⓐ Ⓑ Ⓒ Ⓓ
110. Ⓐ Ⓑ Ⓒ Ⓓ	135. Ⓐ Ⓑ Ⓒ Ⓓ
111. Ⓐ Ⓑ Ⓒ Ⓓ	136. Ⓐ Ⓑ Ⓒ Ⓓ
112. Ⓐ Ⓑ Ⓒ Ⓓ	137. Ⓐ Ⓑ Ⓒ Ⓓ
113. Ⓐ Ⓑ Ⓒ Ⓓ	138. Ⓐ Ⓑ Ⓒ Ⓓ
114. Ⓐ Ⓑ Ⓒ Ⓓ	139. Ⓐ Ⓑ Ⓒ Ⓓ
115. Ⓐ Ⓑ Ⓒ Ⓓ	140. Ⓐ Ⓑ Ⓒ Ⓓ
116. Ⓐ Ⓑ Ⓒ Ⓓ	141. Ⓐ Ⓑ Ⓒ Ⓓ
117. Ⓐ Ⓑ Ⓒ Ⓓ	142. Ⓐ Ⓑ Ⓒ Ⓓ
118. Ⓐ Ⓑ Ⓒ Ⓓ	143. Ⓐ Ⓑ Ⓒ Ⓓ
119. Ⓐ Ⓑ Ⓒ Ⓓ	144. Ⓐ Ⓑ Ⓒ Ⓓ
120. Ⓐ Ⓑ Ⓒ Ⓓ	145. Ⓐ Ⓑ Ⓒ Ⓓ
121. Ⓐ Ⓑ Ⓒ Ⓓ	146. Ⓐ Ⓑ Ⓒ Ⓓ
122. Ⓐ Ⓑ Ⓒ Ⓓ	147. Ⓐ Ⓑ Ⓒ Ⓓ
123. Ⓐ Ⓑ Ⓒ Ⓓ	148. Ⓐ Ⓑ Ⓒ Ⓓ
124. Ⓐ Ⓑ Ⓒ Ⓓ	149. Ⓐ Ⓑ Ⓒ Ⓓ
125. Ⓐ Ⓑ Ⓒ Ⓓ	150. Ⓐ Ⓑ Ⓒ Ⓓ

Exam 2—Morning Session

SITUATION FOR PROBLEMS 101–105

A complete mix activated sludge process has been selected for treatment of a wastewater. Available design information is presented in the following table.

parameter	value
flow rate from primary clarifiers	5000 m³/d
solids recycle flow rate	180 m³/d
mixed liquor volatile suspended solids, MLVSS	2300 mg/L
recycle solids	10 000 mg/L
influent substrate	192 mg/L BOD$_5$
effluent substrate	20 mg/L soluble BOD$_5$
yield coefficient	0.5 at 20°C
growth rate coefficient	0.40/d at 20°C
decay coefficient	0.05/d at 20°C
minimum mean cell residence time	3.0 d
activated sludge safety factor	2.5

101. What is the volume of the complete mix bioreactor?

- (A) 500 m³
- (B) 520 m³
- (C) 900 m³
- (D) 1100 m³

102. What is the daily dry mass of biosolids produced in the bioreactor?

- (A) 310 kg/d
- (B) 320 kg/d
- (C) 430 kg/d
- (D) 450 kg/d

103. What is the daily volume of sludge wasted from the secondary clarifier at 6% solids corresponding to a sludge production rate of 500 kg/d?

- (A) 5.4 m³/d
- (B) 8.3 m³/d
- (C) 19 m³/d
- (D) 27 m³/d

104. What is the specific substrate utilization rate in the bioreactor?

- (A) 0.085 d^{-1}
- (B) 0.19 d^{-1}
- (C) 0.37 d^{-1}
- (D) 0.42 d^{-1}

105. What is the F/M ratio for the bioreactor?

- (A) 0.085 d^{-1}
- (B) 0.19 d^{-1}
- (C) 0.37 d^{-1}
- (D) 0.41 d^{-1}

SITUATION FOR PROBLEMS 106–107

A biological treatment process produces an effluent that requires clarification. The influent solids concentration to the clarifier is 1600 mg/L. The flow rate is 15 000 m³/d, and the clarifier solids flux and settling velocity are 2.3 kg/m²·h and 1.34 m/h, respectively.

106. What is the required total surface area for the clarifier?

- (A) 270 m²
- (B) 360 m²
- (C) 440 m²
- (D) 470 m²

107. What is the required overflow rate for the clarifier if settling velocity control design?

- (A) 1.3 m³/m²·h
- (B) 1.7 m³/m²·h
- (C) 3.0 m³/m²·h
- (D) 9.4 m³/m²·h

SITUATION FOR PROBLEMS 108–111

Bench scale bioreactor tests conducted at 20°C using wastewater samples have provided the following kinetic coefficients.

maximum substrate use rate	5.4 d^{-1}
half-velocity constant	68 mg/L
endogenous decay rate coefficient	0.047 d^{-1}
maximum yield coefficient	0.77

108. What is the value of the maximum specific growth rate at 20°C?

(A) 0.53 d^{-1}
(B) 1.9 d^{-1}
(C) 4.2 d^{-1}
(D) 31 d^{-1}

109. What is the value of the endogenous decay rate coefficient at 15°C if the temperature-activity coefficient is 1.047?

(A) 0.037 d^{-1}
(B) 0.047 d^{-1}
(C) 0.059 d^{-1}
(D) 0.068 d^{-1}

110. What is the value of the substrate utilization rate at 20°C when the effluent substrate concentration is 20 mg/L and the reactor biomass concentration is 2400 mg/L?

(A) −690 mg/L·d
(B) −2900 mg/L·d
(C) −4700 mg/L·d
(D) −32 000 mg/L·d

111. What is the value of the net specific growth rate at 20°C when the maximum specific growth rate is 1 d^{-1} and the effluent substrate concentration is 20 mg/L?

(A) 0.18 d^{-1}
(B) 0.23 d^{-1}
(C) 0.27 d^{-1}
(D) 0.95 d^{-1}

SITUATION FOR PROBLEMS 112-113

A food processing wastewater contains acetic acid as the primary contaminant at a concentration of 0.05M. The wastewater flow is 2500 m³/d.

112. Assuming $C_{172}H_{244}O_{70}N_{35}P_3S$ is a typical chemical formula for a bacterial cell, what are the daily nutrient requirements of N, P, and S to mineralize the acetic acid?

(A) 35 kg/d N, 3 kg/d P, and 1 kg/d S
(B) 350 kg/d N, 66 kg/d P, and 23 kg/d S
(C) 700 kg/d N, 130 kg/d P, and 46 kg/d S
(D) 6300 kg/d N, 24 kg/d P, and 78 kg/d S

113. What is the daily biomass production from the aerobic mineralization of the acetic acid if the equation for the reaction is $13CH_3COO^- + 3NH_4^+ + 11O_2 \rightarrow 3C_5H_7O_2N + 10H_2O + CO_2 + 10HCO_3^-$?

(A) 360 kg/d
(B) 3200 kg/d
(C) 7500 kg/d
(D) 14 000 kg/d

SITUATION FOR PROBLEMS 114-117

Dye tracer studies were conducted to define the flow characteristics of a reaction tank for a water treatment process. The reactor volume was 25,000 gal, and the flow rate to the reactor was 500 gal/min. Water samples collected at the tank effluent following the addition of the tracer provided the following tabulated and graphed results.

time (min)	concentration (μg/L)
20	0
30	100
40	390
50	148
60	83
70	47
80	25
90	12
100	7.5
110	2.5

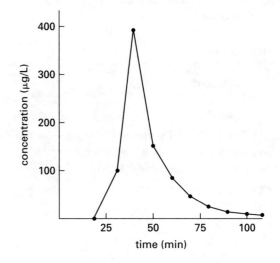

114. What is the reactor mean hydraulic residence time?

(A) 20 min
(B) 47 min
(C) 65 min
(D) 110 min

115. What is the reactor median hydraulic residence time?

(A) 20 min
(B) 38 min
(C) 47 min
(D) 65 min

116. What is the reactor hydraulic residence time that corresponds to the peak of the concentration time plot?

(A) 10 min
(B) 26 min
(C) 40 min
(D) 47 min

117. What is the reactor minimum hydraulic residence time?

(A) $\leq$ 10 min
(B) $\leq$ 20 min
(C) $\leq$ 30 min
(D) $\leq$ 40 min

SITUATION FOR PROBLEMS 118–120

A residential/commercial development is being proposed that would occupy 20 ac and consist of the following facilities.

- 100 unit studio and single bedroom residential low-rise apartments with anticipated average occupancy of two persons per apartment
- building with commercial space to accommodate 100 office employees
- a full-service restaurant that will serve 62 lunch and 62 dinner customers daily and operate 7 day/wk
- a luncheon deli that will serve 36 customers daily during the 5 day business week
- a tennis and swimming club for office employees and apartment residents (40% expected use rate)

Twenty percent of the site will be landscaped and will require irrigation 26 wk/yr with a typical lawn sprinkler rate of 1 in/wk.

118. What is the average annual irrigation water demand?

(A) 2,800,000 gal/yr
(B) 5,700,000 gal/yr
(C) 11,000,000 gal/yr
(D) 14,000,000 gal/yr

119. What is the average annual water demand, excluding irrigation, for the development, assuming typical water consumption rates apply and conventional plumbing fixtures and appliances are used?

(A) 9,000,000 gal/yr
(B) 12,000,000 gal/yr
(C) 19,000,000 gal/yr
(D) 46,000,000 gal/yr

120. If water-conserving plumbing fixtures and appliances are used and landscaping consists of indigenous plants that do not require irrigation, what is the average annual water demand?

(A) 6,000,000 gal/yr
(B) 8,000,000 gal/yr
(C) 10,000,000 gal/yr
(D) 12,000,000 gal/yr

SITUATION FOR PROBLEM 121

The projected average daily water demand schedule for a community water supply is presented in the following table and illustration.

time of day	average flow (m^3/s)
0000 – 0200	0.0126
0200 – 0400	0.0115
0400 – 0600	0.0257
0600 – 0800	0.0720
0800 – 1000	0.0746
1000 – 1200	0.0662
1200 – 1400	0.0573
1400 – 1600	0.0489
1600 – 1800	0.0720
1800 – 2000	0.0400
2000 – 2200	0.0284
2200 – 2400	0.0164

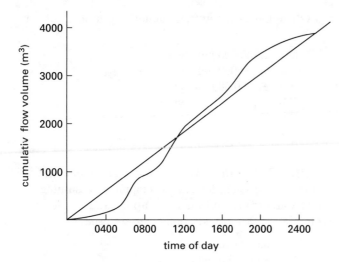

121. What is the reservoir capacity required to meet the average daily demand?

(A) 1000 m³
(B) 2000 m³
(C) 4000 m³
(D) 5000 m³

SITUATION FOR PROBLEMS 122–125

A tank-impellor type flash mixer and a horizontal-paddle-wheel type flocculator are needed to treat a design flow of 5000 m³/d. For the flash mixer, the velocity gradient is 850 s^{-1} with a hydraulic residence time of 120 s. For flocculation, the average velocity gradient over two sections is 45 s^{-1} with a time-velocity gradient of 97 500 (unitless). The flocculator paddle rotational speed is 4 rev/min. The flocculator will be followed by a 3.0 m deep sedimentation basin.

122. What is the total power required at the flash mixer, assuming a single mixing tank is used?

(A) 0.35 kW
(B) 5.0 kW
(C) 17 kW
(D) 30 kW

123. What is the required total volume of the flocculator tank, assuming a single two-section tank is used?

(A) 24 m³
(B) 48 m³
(C) 95 m³
(D) 125 m³

124. What is the total area of the flocculator paddles, assuming a single, two-section, 100 m³ tank is used?

(A) 1.1 m²
(B) 2.1 m²
(C) 3.3 m²
(D) 4.2 m²

125. What is the total power required at the flocculator paddles, assuming a single, two-section, 100 m³ tank is used?

(A) 0.049 kW
(B) 0.097 kW
(C) 0.20 kW
(D) 0.24 kW

SITUATION FOR PROBLEM 126

Two parallel-operated conventional multimedia filters are used to treat a groundwater source prior to chlorination and distribution. The flow rate is 3500 m³/d with other filter characteristics as follows.

filtration rate	5 m³/m²·h
backwash rate	30 m³/m²·h
backwash period	20 min (of every 24 h)
conditioning period	5 min

126. What is the net daily production from each filter?

(A) 1600 m³/d
(B) 1800 m³/d
(C) 3200 m³/d
(D) 3500 m³/d

SITUATION FOR PROBLEM 127

Results of soluble biochemical oxygen demand (BOD) analyses of stream water are as follows. Standardized testing procedures were used.

sample	volume (mL)	DO_1 (mg/L)	DO_5 (mg/L)
1	200	9.1	1.6
2	100	9.2	2.3
3	50	9.3	5.8
4	20	9.3	7.2

The deoxygenation rate constant at 25°C, K_d (base e), is 0.40/d, and the temperature correction coefficient, θ_c, is 1.047. All samples were incubated at 25°C for 5 d.

127. What is the BOD$_5$ concentration at 20°C for the wastewater?

(A) 17 mg/L
(B) 19 mg/L
(C) 21 mg/L
(D) 24 mg/L

SITUATION FOR PROBLEM 128

An old wastewater treatment plant discharges raw sewage during periods of high rainfall. During these bypass periods, the discharge flows are near 2.3 MGD with dissolved oxygen (DO) concentrations near 2.10 mg/L. During these bypass periods, the river receiving the discharge flows at about 100 ft^3/sec with a water temperature of 52°F.

128. What is the DO concentration in the river once complete mixing of the wastewater discharge has occurred?

(A) 2.10 mg/L
(B) 6.60 mg/L
(C) 10.8 mg/L
(D) 11.1 mg/L

SITUATION FOR PROBLEMS 129–131

A wastewater discharge to a river produces the following dissolved oxygen (DO) sag curve.

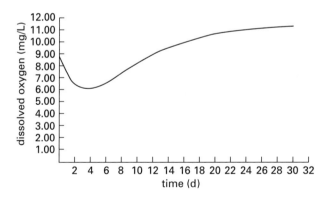

The river receiving the discharge has the following characteristics.

average cross-sectional area	140 ft^2
average flow velocity	0.51 ft/sec
temperature	50°F

129. What is the DO deficit at a location 100 mi downstream of the discharge?

(A) 2.2 mg/L
(B) 3.9 mg/L
(C) 5.2 mg/L
(D) 9.1 mg/L

130. What is the critical DO deficit?

(A) 2.2 mg/L
(B) 3.9 mg/L
(C) 5.2 mg/L
(D) 6.1 mg/L

131. What is the location downstream of the discharge where the maximum DO deficit occurs?

(A) 0 mi
(B) 3.5 mi
(C) 29 mi
(D) 230 mi

SITUATION FOR PROBLEMS 132–133

A pond occupies an area of 24 ac to an average depth of 8 ft. The creek feeding the pond has an average flow of 2.23 ft^3/min with a total Kjeldahl nitrogen (TKN) concentration of 1.43 mg/L as N and an ammonia nitrogen concentration of 0.96 mg/L as N. All organic nitrogen entering the pond is deposited to sediment.

132. What is the annual TKN loading to the pond?

(A) 0.052 lbm/ac-ft-yr
(B) 0.54 lbm/ac-ft-yr
(C) 3.6 lbm/ac-ft-yr
(D) 19 lbm/ac-ft-yr

133. What is the concentration of nitrogen in the pond if all the organic nitrogen is lost to sediment?

(A) 0.47 mg/L
(B) 0.67 mg/L
(C) 0.96 mg/L
(D) 1.4 mg/L

SITUATION FOR PROBLEMS 134–136

A tire manufacturer generates 11 000 kg/d of shredded vulcanized rubber waste from production operations occurring over two 8 h shifts 5 d/wk. The manufacturer also operates an activated sludge plant for production wastewaters. The activated sludge plant produces daily,

7 d/wk, 4 m³ of wasted sludge dewatered to 76% moisture that requires disposal. The heating value and ash content of the waste materials are as follows.

rubber waste	90 000 kJ/kg, 7% ash
wasted sludge	30 000 kJ/kg, 0.5% ash

134. What is the hourly heat value of the shredded rubber?

(A) 1.0×10^7 kJ/h
(B) 2.9×10^7 kJ/h
(C) 4.1×10^7 kJ/h
(D) 5.8×10^7 kJ/h

135. What is the hourly heat value available from the dry sludge solids?

(A) 3.0×10^4 kJ/h
(B) 1.2×10^6 kJ/h
(C) 3.9×10^6 kJ/h
(D) 9.0×10^6 kJ/h

136. How much heat is required to dry the sludge at 100°C from 76% moisture to 0% moisture if the initial temperature of the sludge is 20°C?

(A) 43 000 kJ/h
(B) 240 000 kJ/h
(C) 290 000 kJ/h
(D) 330 000 kJ/h

SITUATION FOR PROBLEMS 137–141

An automotive accessories manufacturer produces a wastewater with the following characteristics.

chemical	concentration (mg/L)
Cr^{+3} (as $Cr_2(SO_4)_3$)	406
Ca^{+2}	187
Mg^{+2}	42
Na^+	115
HCO_3^-	714

Wastewater is discharged from the plating shop at a uniform, continuous rate of 0.7 m³/min during two 8 h shifts, 20 d/mo.

Sodium hydroxide (NaOH) was selected for precipitation of the reduced chromium. The chemical equation for the precipitation reaction is

$$Cr_2(SO_4)_3 + 6NaOH \rightarrow 2Cr(OH)_3 \downarrow + 3Na_2SO_4$$

Caustic soda is available at 73% purity. Sludge is dewatered to 35% solids prior to transportation for disposal.

137. What is the total annual flow volume?

(A) 160 000 m³/yr
(B) 240 000 m³/yr
(C) 330 000 m³/yr
(D) 360 000 m³/yr

138. What is the annual mass of sodium hydroxide at 73% purity required for precipitating the reduced chromium, neglecting any sodium hydroxide demand from hardness?

(A) 13 000 kg/yr
(B) 54 000 kg/yr
(C) 77 000 kg/yr
(D) 130 000 kg/yr

139. What is the annual dry mass of chromium hydroxide sludge produced?

(A) 26 000 kg/yr
(B) 34 000 kg/yr
(C) 68 000 kg/yr
(D) 79 000 kg/yr

140. What is the annual mass of sodium hydroxide at 73% purity required by the precipitation reaction for calcium hardness?

(A) 94 000 kg/y
(B) 100 000 kg/yr
(C) 140 000 kg/yr
(D) 190 000 kg/yr

141. What is the annual dry mass of sludge produced from calcium hardness precipitation?

(A) 16 000 kg/yr
(B) 86 000 kg/yr
(C) 92 000 kg/yr
(D) 170 000 kg/yr

SITUATION FOR PROBLEMS 142–144

Wastewater at 25°C contains cadmium at 121 mg/L in a solution at pH 4.1.

142. What is the concentration of sodium hydroxide required to reduce the cadmium concentration to 2.0 mg/L?

(A) 0.12 mg/L
(B) 0.46 mg/L
(C) 0.69 mg/L
(D) 1.1 mg/L

143. How much sodium hydroxide is required to raise the solution pH to 8.6?

(A) 0.08 mg/L
(B) 0.16 mg/L
(C) 1.0 mg/L
(D) 1.4 mg/L

144. What is the cadmium concentration if the pH is raised to 8.6?

(A) 19 mg/L
(B) 27 mg/L
(C) 37 mg/L
(D) 52 mg/L

SITUATION FOR PROBLEMS 145–148

A leaking waste sump at a computer chip fabrication plant has resulted in contaminated groundwater containing the following mix of chemicals.

mineral spirits	227 ppb
trichloroethene	289 ppb
1,1,1-trichloroethane	292 ppb
isopropanol	170 ppb
methyl ethyl ketone	92 ppb
isophorone	86 ppb
1,2-dichloroethene	183 ppb
1,1-dichloroethane	46 ppb

A bench scale isotherm study was conducted using samples of the groundwater. This study provided the following isotherm equation.

$$X/M = 2.837 C_e^{0.431}$$

X	mass of chemical removed	mg
M	mass of GAC capacity consumed	g
C_e	chemical concentration at equilibrium	mg/L

It is expected that the extraction well system will be able to produce about 92 gal/min of continuous flow. Two adsorber vessels, each with 20,000 lbm granular activated carbon (GAC) capacity, are used in a lead-follow configuration. Assume vessels operate at the same pH and temperature for which the isotherm was developed. The required effluent concentration is 0.1 mg/L.

145. What is the daily mass of chemical removed by the adsorption system?

(A) 0.33 lbm/day
(B) 0.64 lbm/day
(C) 1.4 lbm/day
(D) 3.1 lbm/day

146. What is the GAC adsorption capacity for the chemical mixture?

(A) 0.59 mg/g
(B) 0.84 mg/g
(C) 1.1 mg/g
(D) 3.3 mg/g

147. What is the GAC daily use rate for an adsorption capacity of 1.0 mg/g and an adsorption rate of 1.0 lbm/day?

(A) 430 lbm/day
(B) 1000 lbm/day
(C) 1300 lbm/day
(D) 3300 lbm/day

148. What is the GAC change-out interval per adsorption vessel for a GAC use rate of 500 lbm/day?

(A) 15 days/vessel
(B) 20 days/vessel
(C) 40 days/vessel
(D) 80 days/vessel

SITUATION FOR PROBLEMS 149–150

A community of 37,000 residents produces solid waste at 2.2 lbm/person-day. The trucks available to collect the waste have a 12 yd³ capacity and can compact the waste to 760 lbm/yd³. An idealized collection route within the community has the following characteristics.

time from transfer station to route end/beginning	38 min
time to unload at transfer station	17 min
time to collect waste at each stop	32 sec
time between stops	7 sec
average residents per stop	4

The collection crews work one 8 h shift each day, Monday through Friday.

149. How many loads can one truck complete in a single day?

(A) one load
(B) two loads
(C) three loads
(D) four loads

150. What is the number of trucks required to collect all the waste in the community once weekly if any truck can complete three loads per day?

(A) three trucks
(B) five trucks
(C) seven trucks
(D) nine trucks

Instructions

Name: _____
      Last   First   Middle Initial

Do not enter solutions in the test booklet. Complete solutions must be entered on the answer sheet provided by your proctor.

This is an open-book examination. You may use textbooks, handbooks, and other bound references, along with a battery-operated, silent, nonprinting calculator. Unbound reference materials and notes, scratchpaper, and writing tablets are not permitted. You may not consult with or otherwise share any materials or information with others taking the exam.

You must work all 50 multiple-choice problems in the four-hour period allocated for the afternoon session. Each of the 50 problems is worth 1 point. No partial credit will be awarded. Your score will be based entirely on the responses marked on the answer sheet. You may use blank spaces in the exam booklet for scratch work. However, no credit will be awarded for work shown in margins or on other pages of the exam booklet. Mark only one answer to each problem.

Principles and Practice of Engineering Examination

AFTERNOON SESSION
Sample Examination 2

151. Ⓐ Ⓑ Ⓒ Ⓓ   176. Ⓐ Ⓑ Ⓒ Ⓓ
152. Ⓐ Ⓑ Ⓒ Ⓓ   177. Ⓐ Ⓑ Ⓒ Ⓓ
153. Ⓐ Ⓑ Ⓒ Ⓓ   178. Ⓐ Ⓑ Ⓒ Ⓓ
154. Ⓐ Ⓑ Ⓒ Ⓓ   179. Ⓐ Ⓑ Ⓒ Ⓓ
155. Ⓐ Ⓑ Ⓒ Ⓓ   180. Ⓐ Ⓑ Ⓒ Ⓓ
156. Ⓐ Ⓑ Ⓒ Ⓓ   181. Ⓐ Ⓑ Ⓒ Ⓓ
157. Ⓐ Ⓑ Ⓒ Ⓓ   182. Ⓐ Ⓑ Ⓒ Ⓓ
158. Ⓐ Ⓑ Ⓒ Ⓓ   183. Ⓐ Ⓑ Ⓒ Ⓓ
159. Ⓐ Ⓑ Ⓒ Ⓓ   184. Ⓐ Ⓑ Ⓒ Ⓓ
160. Ⓐ Ⓑ Ⓒ Ⓓ   185. Ⓐ Ⓑ Ⓒ Ⓓ
161. Ⓐ Ⓑ Ⓒ Ⓓ   186. Ⓐ Ⓑ Ⓒ Ⓓ
162. Ⓐ Ⓑ Ⓒ Ⓓ   187. Ⓐ Ⓑ Ⓒ Ⓓ
163. Ⓐ Ⓑ Ⓒ Ⓓ   188. Ⓐ Ⓑ Ⓒ Ⓓ
164. Ⓐ Ⓑ Ⓒ Ⓓ   189. Ⓐ Ⓑ Ⓒ Ⓓ
165. Ⓐ Ⓑ Ⓒ Ⓓ   190. Ⓐ Ⓑ Ⓒ Ⓓ
166. Ⓐ Ⓑ Ⓒ Ⓓ   191. Ⓐ Ⓑ Ⓒ Ⓓ
167. Ⓐ Ⓑ Ⓒ Ⓓ   192. Ⓐ Ⓑ Ⓒ Ⓓ
168. Ⓐ Ⓑ Ⓒ Ⓓ   193. Ⓐ Ⓑ Ⓒ Ⓓ
169. Ⓐ Ⓑ Ⓒ Ⓓ   194. Ⓐ Ⓑ Ⓒ Ⓓ
170. Ⓐ Ⓑ Ⓒ Ⓓ   195. Ⓐ Ⓑ Ⓒ Ⓓ
171. Ⓐ Ⓑ Ⓒ Ⓓ   196. Ⓐ Ⓑ Ⓒ Ⓓ
172. Ⓐ Ⓑ Ⓒ Ⓓ   197. Ⓐ Ⓑ Ⓒ Ⓓ
173. Ⓐ Ⓑ Ⓒ Ⓓ   198. Ⓐ Ⓑ Ⓒ Ⓓ
174. Ⓐ Ⓑ Ⓒ Ⓓ   199. Ⓐ Ⓑ Ⓒ Ⓓ
175. Ⓐ Ⓑ Ⓒ Ⓓ   200. Ⓐ Ⓑ Ⓒ Ⓓ

Exam 2—Afternoon Session

SITUATION FOR PROBLEM 151

Several methods are available to estimate the storage volume for storm water detention basins. Among these is the graphical storage method.

151. Which of the following statements is true regarding the graphical storage method?

(A) It is difficult to use.
(B) It is less accurate than other methods.
(C) It is not compatible with SCS-NRCS runoff calculation methods.
(D) It is best suited for large detention basin design.

SITUATION FOR PROBLEMS 152–153

A landfill receives 224,000 lbm/day, 365 days/yr, of solid and hazardous waste. The waste arrives at the landfill with an average density of 204 lbm/yd^3 and, when placed in its cell, is compacted to 1000 lbm/yd^3. The landfill covers an area of 83 ac and uses 1.0 yd^3 of soil cover for every 5.0 yd^3 of compacted waste. The landfill design capacity is 35 yr.

152. What is the in-place volume of the bottom liner layer of compacted soil if the landfill is a Resource Conservation and Recovery Act (RCRA) Subtitle C facility?

(A) 130,000 yd^3
(B) 200,000 yd^3
(C) 310,000 yd^3
(D) 400,000 yd^3

153. What is the in-place volume of landfilled waste and soil cover at its design capacity?

(A) 570,000 yd^3
(B) 2,300,000 yd^3
(C) 2,900,000 yd^3
(D) 3,400,000 yd^3

SITUATION FOR PROBLEMS 154–155

The major industry in the city generates waste from four different processes. The wastes are characterized in the table for probs. 154–155.

154. What hazardous waste classification criteria are satisfied for waste 1?

(A) ignitability
(B) ignitability and corrosivity
(C) corrosivity and toxicity
(D) toxicity

155. If simple neutralization was applied as treatment for waste 3, would it still meet criteria for classification as hazardous waste?

(A) Yes, because it would still be hazardous based on ignitability.
(B) Yes, because it would still be hazardous based on toxicity.
(C) No, because it was not hazardous based on corrosivity before neutralization.
(D) No, because it was hazardous based on corrosivity only.

Table for Probs. 154–155

characteristic	waste 1	waste 2	waste 3	waste 4
ignitability (flash point, °C)	73	92	136	49
corrosivity (pH)	3.3	12.7	2.0	7.8
reactivity (reactive, yes/no)	no	yes	no	no
toxicity (contaminant, concentration, μg/L)	benzene, 605	–	trichloroethylene, 378	–

SITUATION FOR PROBLEM 56

The U.S. Environmental Protection Agency (EPA) has been criticized for its early approach to a national air-pollution control strategy.

156. What is the traditional approach employed by the EPA for air pollution control?

(A) command-and-control based on uniform emission standards
(B) geographic-specific based on industrial category
(C) variable emission standards based on climatic conditions and topography
(D) site-by-site based on economic and demographic factors

SITUATION FOR PROBLEMS 157–159

A power plant uses coal with a sulfur content of 2.7% at a feed rate of 8.3 lbm/sec. Monitoring at the plant using a hi-vol sampler for particulate matter has produced the following results.

sampling period	24 hr
particle size collected	less than 10 μm
clean filter mass	7.649 g
filter mass at 24 h	7.891 g
initial air flow	50 ft^3/min
final air flow	49.3 ft^3/min

Observations of the plume at the stack show a 35% opacity.

157. What is the daily sulfur (as SO_2) emitted to air pollution control (APC) equipment if 5% of the sulfur goes to each?

(A) 1900 lbm/day
(B) 9200 lbm/day
(C) 18,000 lbm/day
(D) 37,000 lbm/day

158. What is the PM_{10} concentration at the sampling station?

(A) 48 μg/m^3
(B) 120 μg/m^3
(C) 210 μg/m^3
(D) 240 μg/m^3

159. What is the Ringlemann number corresponding to the opacity?

(A) $1^1/_4$
(B) $1^3/_4$
(C) 2
(D) $2^1/_4$

SITUATION FOR PROBLEMS 160–161

The following problems address secondary air pollutants.

160. What emitted sulfur and nitrogen compounds most often occur as precursors to secondary pollutants?

(A) nitric oxide and elemental sulfur
(B) nitric oxide and sulfur dioxide
(C) nitrogen dioxide and sulfur dioxide
(D) nitrogen dioxide and sulfur trioxide

161. What are the primary nuisance components of photochemical smog?

(A) particulates and sulfur oxides
(B) carbon dioxide and CFCs
(C) ozone and aldehydes
(D) nitrogen oxides and carbon monoxide

SITUATION FOR PROBLEMS 162–165

A cross-current flow scrubber has been selected for the air pollution control (APC) process at a manufacturing facility. The air flow rate to be treated is 10 m^3/s. The cross-current flow scrubber is defined by the following parameters.

gas velocity	34 cm/s
liquid-gas flow ratio	0.6×10^{-3}
fractional target efficiency	0.052
scrubber contact zone length	4.25 m
liquid droplet diameter	0.028 cm
spray nozzle pressure	0.05 atm
air viscosity	2.5×10^{-4} g/cm·s $= 2.5 \times 10^{-5}$ kg/m·s
particulate concentration	20 g/m^3
particle diameter	4.0 μm

162. What is the efficiency of the scrubber?

(A) 13%
(B) 49%
(C) 51%
(D) 87%

163. What is the required cross-sectional area for the scrubber?

(A) 3.4 m²
(B) 29 m²
(C) 294 m²
(D) 340 m²

164. What is the air pressure drop through the scrubber spray nozzle?

(A) 0.005 atm
(B) 0.05 atm
(C) 0.5 atm
(D) 1.0 atm

165. How will increasing the cross-sectional area influence the scrubber efficiency?

(A) The efficiency will increase because the scrubber volume will increase.
(B) The efficiency will increase because the gas velocity, v_g, will increase.
(C) The efficiency will decrease because the gas velocity, v_g, will decrease.
(D) The efficiency will decrease because liquid-gas flow ratio will decrease.

SITUATION FOR PROBLEMS 166–167

Hydrogen cyanide gas is recovered from an air stream using a counter-current scrubber operated under vacuum. The gas flow rate through the scrubber is 1.9 m³/s with a hydrogen cyanide concentration of 1.94 g/m³. The scrubbing solution is 0.1 N sodium hydroxide.

166. What is the required flow rate for the sodium hydroxide solution?

(A) 22 L/min
(B) 41 L/min
(C) 55 L/min
(D) 82 L/min

167. For a sodium hydroxide solution flow rate of 50 L/min, what is the resulting concentration of sodium cyanide?

(A) 0.070 N
(B) 0.12 N
(C) 0.16 N
(D) 0.26 N

SITUATION FOR PROBLEMS 168–170

A pulsed-air baghouse is used to control particulate emissions from a cosmetics manufacturing facility. The gas stream characteristics are

flow rate	18 m³/s
particulate concentration	21 g/m³

The baghouse characteristics are

filtering velocity	0.08 m³/m²·s
fabric resistance	0.32 kPa/(cm/s)
cake resistance	0.98 kPa/(cm/s)(g/cm²)
bag diameter	20 cm
bag length	9 m

168. Will a 5 min cleaning interval be appropriate, if an allowable pressure drop of between 1.5 to 2.0 kPa is acceptable?

(A) Yes, it is within the typical range.
(B) Yes, it does not exceed the maximum recommended value.
(C) No, it is too short.
(D) No, it is too long.

169. How many bags are required?

(A) 40 bags
(B) 140 bags
(C) 800 bags
(D) 3200 bags

170. How many compartments are typical for a total fabric area of 1500 m²?

(A) 7 compartments
(B) 10 compartments
(C) 13 compartments
(D) 16 compartments

SITUATION FOR PROBLEMS 171–172

The following illustration represents atmospheric stability conditions at a site with a ground surface elevation of 100 m.

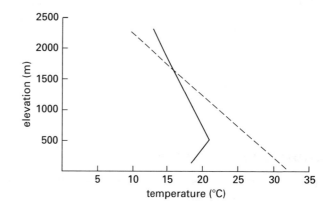

171. Which phrase best describes the atmospheric conditions?

- (A) inversion to 500 m, subadiabatic above 500 m
- (B) inversion to 500 m, superadiabatic above 500 m
- (C) subadiabatic to 500 m, superadiabatic above 500 m
- (D) superadiabatic to 500 m, subadiabatic above 500 m

172. What is the depth of the maximum mixing zone?

- (A) 500 m
- (B) 1100 m
- (C) 1500 m
- (D) 1600 m

SITUATION FOR PROBLEMS 173-175

A stack with an effective height of 120 m emits pollutants at 5.4 kg/s. Wind speed measured 10 m above the ground surface is 2.6 m/s on a sunny day with moderate incoming solar radiation.

173. How far downwind is the maximum ground-level concentration of pollutants emitted from the stack observed?

- (A) 420 m
- (B) 800 m
- (C) 1600 m
- (D) 5000 m

174. For pollutants emitted from the stack, what is the ground-level concentration along the plume centerline at a distance 2000 m downwind of the source?

- (A) 7.2 mg/m^3
- (B) 8.7 mg/m^3
- (C) 9.8 mg/m^3
- (D) 27 mg/m^3

175. For pollutants emitted from the stack, what is the ground-level concentration at a distance 1600 m downwind and 300 m crosswind of the source?

- (A) 3.5 mg/m^3
- (B) 6.0 mg/m^3
- (C) 6.6 mg/m^3
- (D) 16 mg/m^3

SITUATION FOR PROBLEM 76

Atmospheric conditions include an inversion with a 520 m base and category D stability.

176. At what distance downwind of an 80 m stack does the inversion form of the dispersion equation apply?

- (A) 20 km
- (B) 30 km
- (C) 40 km
- (D) 60 km

SITUATION FOR PROBLEMS 177-181

Arsenic at 270 µg/L has been discovered in a spring serving a rural community with a population of 5000. The spring feeds a small creek that provides the community with a trout fishery. The community has used the spring as a drinking water source for 12 yr, but it is unknown how long the contamination has existed. The toxicology of arsenic is characterized by the following parameters

oral route potency factor (PF)	1.75 (mg/kg·d)$^{-1}$
inhalation route potency factor (PF)	50 (mg/kg·d)$^{-1}$
reference dose (RfD)	0.0003 mg/kg·d
bioconcentration factor (BCF)	44 L/kg

The exposure factors representing the community are

ingestion (drinking water)	2 L/d (adult)
	1 L/d (child)
ingestion (fish)	54 g/d
inhalation	20 m^3/d (adult)
	12 m^3/d (child)
lifespan	70 yr
body mass	70 kg (adult)
	15 kg (child)

177. What is the lifetime cancer risk from arsenic in the drinking water for an adult living in the exposed community?

- (A) 1.9×10^{-4}
- (B) 2.3×10^{-3}
- (C) 1.3×10^{-2}
- (D) 7.5×10^{-1}

178. What is the lifetime cancer risk for adult residents exposed to arsenic through inhalation, if the exposure duration is 30 min daily and the concentration is 0.13 µg/m^3?

- (A) 5.4×10^{-7}
- (B) 6.6×10^{-6}
- (C) 3.8×10^{-5}
- (D) 3.1×10^{-4}

179. What is the hazard quotient for a child exposed to the arsenic through drinking water?

(A) 13
(B) 26
(C) 60
(D) 120

180. What is the incremental lifetime cancer risk for an adult who eats fish caught in the creek?

(A) 9.1×10^{-4}
(B) 2.8×10^{-3}
(C) 5.3×10^{-3}
(D) 1.6×10^{-2}

181. What is the cumulative cancer risk for an adult in the community who experiences incremental risks from ingestion of drinking water, ingestion of fish, and inhalation of 9.2×10^{-5}, 1.7×10^{-7}, and 1.8×10^{-4}, respectively?

(A) 9.0×10^{-9}
(B) 1.7×10^{-7}
(C) 2.7×10^{-4}
(D) 1.8×10^{-4}

SITUATION FOR PROBLEMS 182–186

The following problems pertain to material safety data sheets (MSDS).

182. What criteria define a chemical included on a MSDS as hazardous?

(A) Exposure at elevated concentrations for periods greater than 30 min causes dizziness.
(B) The material is a noncarcinogen, but is volatile at ambient temperature and pressure.
(C) The chemical is specifically included on the Z list (29 CFR 1910, Subpart Z).
(D) No threshold limit value has been assigned for the chemical.

183. What worker protection measures would be described on a MSDS?

(A) workplace ventilation requirements
(B) personal breathing apparatus requirements
(C) personal protective clothing requirements
(D) all of the above

184. What information does the heading of a MSDS provide?

(A) the name, address, and contact information for the material manufacturer or supplier
(B) the name of the material covered by the MSDS
(C) the date that the MSDS was issued or revised
(D) all of the above

185. When does the material name on the MSDS have to exactly match the name printed on the material container?

(A) only when the material will be used in a hospital or other health care facility
(B) only when the material is shipped in containers with capacity greater than 30 L or 25 kg
(C) only when the container is labeled with a NFPA hazard diamond
(D) always

186. For mixtures of chemicals, what information does the MSDS provide regarding ingredients?

(A) a complete list of all chemicals included in the mixture, regardless of their hazardous properties
(B) a complete list of all hazardous materials included in the mixture
(C) a list of hazardous materials included in the mixture only if present at greater than 2% of the total mixture
(D) a list of carcinogenic materials only.

SITUATION FOR PROBLEMS 187–188

The following problems relate to workplace exposure to air contaminants.

187. When is providing protective equipment to prevent employee exposure to air contaminants an acceptable control measure?

(A) anytime a potential exposure exists
(B) only when administrative and engineering controls cannot be implemented
(C) only when exposure will exceed the TWA PEL for 2 hr or more during an 8 hr work shift
(D) only when exposure will exceed the TWA PEL for 8 hr or more during a 40 hr workweek

188. When may an employee be exposed to an air contaminant listed with an acceptable ceiling concentration and maximum peak concentration and duration?

(A) only when the exposure period does not exceed the listed duration at any concentration

(B) only when the exposure does not exceed the acceptable ceiling concentration and the exposure period is less than the listed duration

(C) only when the exposure does not exceed the maximum peak concentration for a period greater than the listed duration

(D) only when the exposure does not exceed the maximum peak concentration and the exposure period is less than the listed duration

SITUATION FOR PROBLEMS 189-193

An underground storage tank has leaked benzene into an unconfined aquifer over a period of several months.

For benzene, the organic carbon partition coefficient is 91 mL/g.

The aquifer characteristics are

ambient temperature	8°C
hydraulic conductivity	123 ft/day
hydraulic gradient	0.00091 ft/ft
effective porosity	0.31
total organic carbon	198 mg/kg
bulk density	1.65 g/cm^3

189. What is the average velocity of the groundwater?

(A) 0.033 m/d
(B) 0.11 m/d
(C) 0.44 m/d
(D) 80 m/d

190. What is the average velocity of benzene relative to the groundwater velocity, v_{aw}?

(A) $0.033v_{aw}$
(B) $0.096v_{aw}$
(C) $0.22v_{aw}$
(D) $0.91v_{aw}$

191. For a distance of 100 m from the release point, what is the longitudinal dynamic dispersivity?

(A) 0.27
(B) 1.8
(C) 2.3
(D) 15

192. For a longitudinal dynamic dispersivity of 1.0, what is the longitudinal hydrodynamic dispersion based on a benzene velocity of 1.0 m/d?

(A) 0.034 m^2/d
(B) 0.31 m^2/d
(C) 1.0 m^2/d
(D) 38 m^2/d

193. What is the benzene concentration 90 d after initial release at a distance of 100 m from the release point if the initial concentration was 1000 µg/L, the longitudinal hydrodynamic dispersion is 1.0 m^2/d, and the benzene velocity is 1.0 m/d ?

(A) 0.93 µg/L
(B) 86 µg/L
(C) 230 µg/L
(D) 500 µg/L

SITUATION FOR PROBLEM 94

The following problem addresses degradation sequences of organic chemicals.

194. In which of the following sequences of organic chemicals would vinyl chloride most likely occur next?

(A) carbon tetrachloride, chloroform, methylene chloride

(B) chloroform, methylene chloride, methyl chloride

(C) perchloroethylene, trichloroethylene, dichloroethylene

(D) tetrachloroethane, trichloroethane, dichloroethane

SITUATION FOR PROBLEMS 195-197

Problems 95 through 97 address quantification and characterization of malodors.

195. At what concentration in the air can the human olfactory sense detect common malodors associated with municipal sewage?

(A) 0.0005 ppm
(B) 0.005 ppm
(C) 0.05 ppm
(D) 0.5 ppm

196. What method is commonly used to characterize and measure odors?

(A) correlation with concentration of conventional parameters such as biochemical oxygen demand (BOD) and total suspended solids (TSS)
(B) organoleptic or sensory detection by humans
(C) air sampling and laboratory analyses
(D) none of the above

197. What is the threshold odor number (TON) where 25 mL of wastewater sample is diluted to 200 mL with odor-free water to produce a barely perceptible odor?

(A) 0.13
(B) 1.1
(C) 8.0
(D) 25

SITUATION FOR PROBLEMS 198–200

The following problems pertain to radon exposure and exposure mitigation.

198. Which of the following is not a measure for mitigating radon in residential structures?

(A) block wall depressurization
(B) crawlspace depressurization
(C) attic depressurization
(D) sub-slab depressurization

199. How is radon detected inside buildings?

(A) by certified professionals using specialized analytical equipment
(B) by individuals using a simple test kit purchased over-the-counter at local retail outlets
(C) by permanently installed active monitoring devices
(D) by all of the above

200. Are homes a more likely source of radon exposure than other buildings?

(A) Yes, homes are constructed of masonry products that are more likely to contain radon.
(B) Yes, people spend the largest percentage of their time in their homes.
(C) Yes, homes are more likely than other structures to have basements.
(D) No, the likelihood of radon exposure is not greater in homes than in other buildings.

Exam 2—Solutions

101. D	126. A	151. B	176. C
102. A	127. B	152. D	177. B
103. B	128. C	153. D	178. B
104. C	129. A	154. D	179. C
105. D	130. C	155. D	180. B
106. D	131. C	156. A	181. C
107. A	132. B	157. D	182. D
108. C	133. C	158. B	183. D
109. A	134. B	159. B	184. D
110. B	135. B	160. B	185. D
111. A	136. D	161. C	186. B
112. C	137. A	162. C	187. B
113. B	138. D	163. B	188. D
114. B	139. C	164. C	189. B
115. B	140. D	165. C	190. D
116. C	141. D	166. D	191. D
117. C	142. C	167. C	192. C
118. A	143. B	168. D	193. B
119. B	144. C	169. A	194. C
120. C	145. C	170. B	195. A
121. A	146. D	171. A	196. B
122. B	147. B	172. C	197. C
123. D	148. C	173. B	198. C
124. C	149. B	174. B	199. D
125. C	150. B	175. B	200. B

101. The mean cell residence time for design is

$$\theta_c = \theta_{c,\min}(\text{SF}) = (3.0 \text{ d})(2.5)$$
$$= 7.5 \text{ d}$$

The hydraulic residence time in the bioreactor is given as follows.

$$\frac{1}{\theta_c} = \frac{Y(S_o - S)}{\theta X} - k_d$$

$$\frac{1}{7.5 \text{ d}} = \frac{(0.5)\left(192 \frac{\text{mg}}{\text{L}} - 20 \frac{\text{mg}}{\text{L}}\right)}{\theta\left(2300 \frac{\text{mg}}{\text{L}}\right)} - \frac{0.05}{\text{d}}$$

$$\theta = 0.204 \text{ d}$$

The bioreactor volume is

$$V = (Q + Q_R)\theta$$
$$= \left(5000 \frac{\text{m}^3}{\text{d}} + 180 \frac{\text{m}^3}{\text{d}}\right)(0.204 \text{ d})$$
$$= 1057 \text{ m}^3 \quad (1100 \text{ m}^3)$$

The answer is (D).

102. The mean cell residence time for design is

$$\theta_c = \theta_{c,\min}(\text{SF}) = (3.0 \text{ d})(2.5)$$
$$= 7.5 \text{ d}$$

$$Y_{\text{obs}} = \frac{Y}{1 + k_d\theta_c} = \frac{0.5}{1 + \left(\frac{0.05}{\text{d}}\right)(7.5 \text{ d})}$$
$$= 0.36$$

The biosolids production is

$$Y_{\text{obs}}(S_o - S)Q = (0.36)\left(192 \frac{\text{mg}}{\text{L}} - 20 \frac{\text{mg}}{\text{L}}\right)$$
$$\times \left(5000 \frac{\text{m}^3}{\text{d}}\right)\left(10^{-3} \frac{\text{kg·L}}{\text{mg·m}^3}\right)$$
$$= 310 \text{ kg/d}$$

The answer is (A).

103. Assume the wasted wet sludge density is approximately that of water at 1000 kg/m³.

The volume of wasted sludge at 6% solids is

$$\frac{500 \frac{\text{kg}}{\text{d}}}{\left(1000 \frac{\text{kg}}{\text{m}^3}\right)(0.06)} = 8.3 \text{ m}^3/\text{d}$$

The answer is (B).

104. The mean cell residence time for design is

$$\theta_c = \theta_{c,\min}(\text{SF}) = (3.0 \text{ d})(2.5)$$
$$= 7.5 \text{ d}$$

$$\frac{1}{\theta_c} = \frac{Y(S_o - S)}{\theta X} - k_d$$

$$\frac{1}{7.5 \text{ d}} = \frac{(0.5)\left(192 \frac{\text{mg}}{\text{L}} - 20 \frac{\text{mg}}{\text{L}}\right)}{\theta\left(2300 \frac{\text{mg}}{\text{L}}\right)} - \frac{0.05}{\text{d}}$$

$$\theta = 0.204 \text{ d}$$

$$U = \frac{S_o - S}{\theta X} = \frac{192 \frac{\text{mg}}{\text{L}} - 20 \frac{\text{mg}}{\text{L}}}{(0.204 \text{ d})\left(2300 \frac{\text{mg}}{\text{L}}\right)}$$
$$= 0.37 \text{ d}^{-1}$$

The answer is (C).

105. The mean cell residence time for design is

$$\theta_c = \theta_{c,\min}(\text{SF}) = (3.0 \text{ d})(2.5)$$
$$= 7.5 \text{ d}$$

$$\frac{1}{\theta_c} = \frac{Y(S_o - S)}{\theta X} - k_d$$

$$\frac{1}{7.5 \text{ d}} = \frac{(0.5)\left(192 \frac{\text{mg}}{\text{L}} - 20 \frac{\text{mg}}{\text{L}}\right)}{\theta\left(2300 \frac{\text{mg}}{\text{L}}\right)} - \frac{0.05}{\text{d}}$$

$$\theta = 0.204 \text{ d}$$

$$\frac{F}{M} = \frac{S_o}{\theta X} = \frac{192 \frac{\text{mg}}{\text{L}}}{(0.204 \text{ d})\left(2300 \frac{\text{mg}}{\text{L}}\right)}$$
$$= 0.41 \text{ d}^{-1}$$

The answer is (D).

106. Based on solids flux,

$$A_s = \frac{QX}{G}$$

$$\frac{\left(15\,000 \frac{\text{m}^3}{\text{d}}\right)\left(1600 \frac{\text{mg}}{\text{L}}\right)\left(10^{-3} \frac{\text{kg·L}}{\text{mg·m}^3}\right)}{\left(2.3 \frac{\text{kg}}{\text{m}^2 \cdot \text{h}}\right)\left(24 \frac{\text{h}}{\text{d}}\right)} = 435 \text{ m}^2$$

Based on settling velocity,

$$A_s = \frac{Q}{v_s}$$

$$\frac{15\,000 \frac{\text{m}^3}{\text{d}}}{\left(1.34 \frac{\text{m}}{\text{h}}\right)\left(24 \frac{\text{h}}{\text{d}}\right)} = 466 \text{ m}^2$$

Since 466 m² > 435 m², settling velocity controls. The required total surface area is

$$A_s = 466 \text{ m}^2 \quad (470 \text{ m}^2)$$

The answer is (D).

107. When settling velocity controls design,

$$v_s = q_o = \frac{Q}{A_s}$$

The required overflow rate is

$$q_o = 1.34 \frac{\text{m}}{\text{h}}$$
$$= 1.34 \text{ m}^3/\text{m}^2 \cdot \text{h} \quad (1.3 \text{ m}^3/\text{m}^2 \cdot \text{h})$$

The answer is (A).

108.
$$k = \frac{\mu_{max}}{Y}$$
$$\mu_{max} = kY = \left(\frac{5.4}{\text{d}}\right)(0.77)$$
$$= 4.158 \text{ d}^{-1} \quad (4.2 \text{ d}^{-1})$$

The answer is (C).

109. The value of the endogenous decay rate coefficient is

$$k_{d15} = k_{d20}\theta^{15-20} = \left(\frac{0.047}{\text{d}}\right)(1.047)^{15-20}$$
$$= 0.037 \text{ d}^{-1}$$

The answer is (A).

110. The substrate utilization rate is

$$r_{su} = \frac{-kXS}{K_s + S} = \frac{-\left(\frac{5.4}{\text{d}}\right)\left(2400 \frac{\text{mg}}{\text{L}}\right)\left(20 \frac{\text{mg}}{\text{L}}\right)}{68 \frac{\text{mg}}{\text{L}} + 20 \frac{\text{mg}}{\text{L}}}$$
$$= -2945 \text{ mg/L} \cdot \text{d} \quad (-2900 \text{ mg/L} \cdot \text{d})$$

The answer is (B).

111.

$$\mu = \mu_{max}\left(\frac{S}{K_s + S}\right) = \frac{\left(\frac{1.0}{\text{d}}\right)\left(20 \frac{\text{mg}}{\text{L}}\right)}{68 \frac{\text{mg}}{\text{L}} + 20 \frac{\text{mg}}{\text{L}}}$$
$$= 0.227 \text{ d}^{-1}$$

The net specific growth rate is

$$\mu' = \mu - k_d = \frac{0.227}{\text{d}} - \frac{0.047}{\text{d}}$$
$$= 0.18 \text{ d}^{-1}$$

The answer is (A).

112. The cell ratio of carbon (C) to nitrogen (N) to phosphorous (P) to sulfur (S) is 172 to 35 to 3 to 1. Acetic acid is CH_3COOH, or $C_2H_4O_2$. In 1 mol acetic acid there are $(2)(0.05)$ mol/L C, or 0.10 mol/L C.

$$\frac{0.10}{172} = \frac{\text{N}}{35} = \frac{\text{P}}{3} = \frac{\text{S}}{1}$$

$$\text{N} = \frac{(35)\left(0.10 \frac{\text{mol}}{\text{L}}\right)}{172} = 0.020 \text{ mol/L N}$$

$$\text{P} = \frac{(3)\left(0.10 \frac{\text{mol}}{\text{L}}\right)}{172} = 0.0017 \text{ mol/L P}$$

$$\text{S} = \frac{(1)\left(0.10 \frac{\text{mol}}{\text{L}}\right)}{172} = 0.00058 \text{ mol/L S}$$

$$\left(0.020 \frac{\text{mol}}{\text{L}} \text{ N}\right)\left(14 \frac{\text{g}}{\text{mol}}\right)$$
$$\times \left(2500 \frac{\text{m}^3}{\text{d}}\right)\frac{(1 \text{ kg})(1000 \text{ L})}{(1000 \text{ g})(1 \text{ m}^3)} = 700 \text{ kg/d N}$$

$$\left(0.0017 \frac{\text{mol}}{\text{L}} \text{ P}\right)\left(31 \frac{\text{g}}{\text{mol}}\right)$$
$$\times \left(2500 \frac{\text{m}^3}{\text{d}}\right)\frac{(1 \text{ kg})(1000 \text{ L})}{(1000 \text{ g})(1 \text{ m}^3)} = 132 \text{ kg/d P}$$
$$(130 \text{ kg/d P})$$

$$\left(0.00058 \frac{\text{mol}}{\text{L}} \text{ S}\right)\left(32 \frac{\text{g}}{\text{mol}}\right)$$
$$\times \left(2500 \frac{\text{m}^3}{\text{d}}\right)\frac{(1 \text{ kg})(1000 \text{ L})}{(1000 \text{ g})(1 \text{ m}^3)} = 46 \text{ kg/d S}$$

The answer is (C).

113. Biomass is represented by $C_5H_7O_2N$. There are 3 mol $C_5H_7O_2N$ per 13 mol CH_3COO^-.

$$\frac{0.05 \frac{\text{mol}}{\text{L}} CH_3COO^-}{13 \text{ mol } CH_3COO^-} = \frac{[C_5H_7O_2N]}{3 \text{ mol } C_5H_7O_2N}$$
$$[C_5H_7O_2N] = 0.0115 \text{ mol/L}$$

$$C_5H_7O_2N \text{ MW} = (5)\left(12 \frac{\text{g}}{\text{mol}}\right) + (7)\left(1 \frac{\text{g}}{\text{mol}}\right)$$
$$+ (2)\left(16 \frac{\text{g}}{\text{mol}}\right) + 14 \frac{\text{g}}{\text{mol}}$$
$$= 113 \text{ g/mol}$$

$$\dot{m} = [C_5H_7O_2N](C_5H_7O_2N \text{ MW})Q$$

$$\left(0.0115\ \frac{\text{mol}}{\text{L}}\right)\left(113\ \frac{\text{g}}{\text{mol}}\right)$$
$$\times \left(2500\ \frac{\text{m}^3}{\text{d}}\right)\frac{(1\ \text{kg})(1000\ \text{L})}{(1000\ \text{g})(1\ \text{m}^3)} = 3249\ \text{kg/d}$$
$$(3200\ \text{kg/d})$$

The answer is (B).

114.

t_i (min)	C_i (μg/L)	$C_i \Delta t$ (min·μg/L)	$C_i \Delta t t_i$ (min^2·μg/L)
20	0	0	0
30	100	1000	30 000
40	390	3900	156 000
50	148	1480	74 000
60	83	830	49 800
70	47	470	32 900
80	25	250	20 000
90	12	120	10 800
100	7.5	75	7500
110	2.5	25	2750
		8150	383 750

C_i tracer concentration at t_i μg/L
Δt time interval between time measurements   min

$$\Delta t = 10\ \text{min}$$

$$t_a = \frac{\Sigma C_i \Delta t t_i}{\Sigma C_i \Delta t} = \frac{383\,750\ \frac{\text{min}^2 \cdot \mu\text{g}}{\text{L}}}{8150\ \frac{\text{min}\cdot\mu\text{g}}{\text{L}}}$$
$$= 47\ \text{min}$$

The answer is (B).

115. t_{med} = time corresponding to $\Sigma C_i \Delta t/2$.

t_i (min)	C_i (μg/L)	$C_i \Delta t$ (min·μg/L)	cumulative $C_i \Delta t$ (min·μg/L)
20	0	0	0
30	100	1000	1000
40	390	3900	4900
50	148	1480	6380
60	83	830	7210
70	47	470	7680
80	25	250	7930
90	12	120	8050
100	7.5	75	8125
110	2.5	25	8150
		8150	

$$\frac{\Sigma C_i \Delta t}{2} = \frac{8150\ \frac{\text{min}\cdot\mu\text{g}}{\text{L}}}{2} = 4075\ \text{min}\cdot\mu\text{g/L}$$

The corresponding time is 38 min.
$$t_{\text{med}} = 38\ \text{min}$$

The answer is (B).

116. Let t_m be the time corresponding to the peak of the concentration time plot.

From the illustration, $t_m = 40$ min.

The answer is (C).

117. Let t_{min} be the time corresponding to the first detectable concentration.
$$t_{\text{min}} \le 30\ \text{min}$$

The answer is (C).

118. The average annual irrigation water demand is
$$(0.20)(20\text{ac})\left(26\ \frac{\text{wk}}{\text{yr}}\right)\left(1\ \frac{\text{in}}{\text{wk}}\right)\left(\frac{1\ \text{ft}}{12\ \text{in}}\right)$$
$$\times \left(43{,}560\ \frac{\text{ft}^2}{\text{ac}}\right)\left(7.48\ \frac{\text{gal}}{\text{ft}^3}\right)$$
$$= 2{,}823{,}850\ \text{gal/yr}\ \ (2{,}800{,}000\ \text{gal/yr})$$

The answer is (A).

119.

use	people	gal/day	days/wk	gal/wk
apartment	200	100	7	140,000
office	100	15	5	7500
restaurant	(2)(62)	9	7	7812
deli	36	6	5	1080
club	(0.4)(200)	100	7	56,000
	(0.4)(100)	100	5	20,000
				232,392

$$\left(232{,}392\ \frac{\text{gal}}{\text{wk}}\right)\left(52\ \frac{\text{wk}}{\text{yr}}\right) = 12{,}084{,}384\ \text{gal/yr}$$
$$(12{,}000{,}000\ \text{gal/yr})$$

The answer is (B).

120. Water conservation devices provide from 13% to 22% use reduction. The average reduction is approximately 18%. The conserved use is 82% of normal.

use	people	gal/day	days/wk	gal/wk
apartment	200	(0.82)(100)	7	114,800
office	100	(0.82)(15)	5	6150
restaurant	(2)(62)	(0.82)(9)	7	6406
deli	36	(0.82)(6)	5	886
club	(0.4)(200)	(0.82)(100)	7	45,920
	(0.4)(100)	(0.82)(100)	5	16,400
				190,562

$$\left(190{,}562 \ \frac{\text{gal}}{\text{wk}}\right)\left(52 \ \frac{\text{wk}}{\text{yr}}\right) = 9{,}909{,}224 \ \text{gal/yr}$$

$$(10{,}000{,}000 \ \text{gal/yr})$$

The answer is (C).

121.

flow volume (m^3/period)	cumulative flow volume (m^3)
91	91
83	174
185	359
518	877
537	1414
477	1891
413	2304
352	2656
518	3174
288	3462
204	3666
118	3784
3784	

$$\text{cumulative flow volume} = (\text{flow period})(\text{average flow})$$

$$= \left(2 \ \frac{\text{h}}{\text{period}}\right)\left(0.0126 \ \frac{\text{m}^3}{\text{s}}\right)\left(3600 \ \frac{\text{s}}{\text{h}}\right)$$

$$= 91 \ \frac{\text{m}^3}{\text{period}}$$

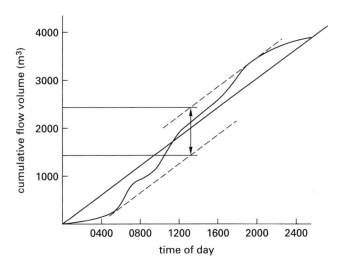

The volume required for storage is 1000 m^3.

The answer is (A).

122.
$$G^2 = \frac{P}{V\mu}$$
$$P = G^2 V \mu$$

The water viscosity is $\mu = 0.001002$ N·s/m^2 at 20°C assumed temperature.

$$V = Qt = \left(5000 \ \frac{\text{m}^3}{\text{d}}\right)(120 \ \text{s})\left(\frac{1 \ \text{d}}{86\,400 \ \text{s}}\right)$$
$$= 6.94 \ \text{m}^3$$

$$P = \left(\frac{850}{\text{s}}\right)^2 (6.94 \ \text{m}^3)\left(0.001002 \ \frac{\text{N·s}}{\text{m}^2}\right)$$
$$\times \left(\frac{1 \ \text{kW·s}}{1000 \ \text{N·m}}\right)$$
$$= 5.0 \ \text{kW}$$

The answer is (B).

123.

$$t = \frac{Gt}{G} = \frac{97\,500}{\left(\frac{45}{\text{s}}\right)\left(60 \ \frac{\text{s}}{\text{min}}\right)}$$
$$= 36 \ \text{min}$$

$$V = \left(5000 \ \frac{\text{m}^3}{\text{d}}\right)(36 \ \text{min})\left(\frac{1 \ \text{d}}{1440 \ \text{min}}\right)$$
$$= 125 \ \text{m}^3$$

The answer is (D).

124. Assume $\rho_w = 1000$ kg/m^3 and $C_D = 1.8$ for flat paddles.

Select the flocculator depth to match the sedimentation basin depth of 3.0 m.

Select the flocculator paddle wheel diameter to rotate on a horizontal axis with a 0.2 m clearance between the paddle tip and the bottom of the tank.

The paddle wheel diameter is
$$3 \ \text{m} - (0.2 \ \text{m})(2) = 2.6 \ \text{m}$$

$$\omega = \frac{\text{v}_{pa}}{\pi D} = 4 \ \text{rev/min}$$

$$\text{v}_{pa} = \omega(0.75) \quad \left[\begin{array}{c}\text{where 75\% slippage exists between}\\ \text{the paddle and the water}\end{array}\right]$$

$$= \left(2.6\pi \ \frac{\text{m}}{\text{rev}}\right)(0.75)\left(4 \ \frac{\text{rev}}{\text{min}}\right)\left(\frac{1 \ \text{min}}{60 \ \text{sec}}\right)$$
$$= 0.408 \ \text{m/s}$$

$$A_p = \frac{G^2 2 V \mu}{C_D \rho_w \text{v}_p^3}$$

$$= \frac{\left(\frac{45}{\text{s}}\right)^2 (2)(100 \ \text{m}^3)\left(0.001002 \ \frac{\text{kg}}{\text{m·s}}\right)}{(1.8)\left(1000 \ \frac{\text{kg}}{\text{m}^3}\right)\left(0.408 \ \frac{\text{m}}{\text{s}}\right)^3}$$

$$= 3.3 \ \text{m}^2$$

The answer is (C).

125.
$$P = G^2 V \mu$$
$$= \left(\frac{45}{\text{s}}\right)^2 (100 \text{ m}^3) \left(0.001002 \frac{\text{N·s}}{\text{m}^2}\right)$$
$$\times \left(\frac{1 \text{ kW·s}}{1000 \text{ N·m}}\right)$$
$$= 0.203 \text{ kW} \quad (0.20 \text{ kW})$$

The answer is (C).

126.
$$V_s = \frac{Q}{A_s}$$
$$A_s = \frac{Q}{V_s} = \frac{\left(3500 \frac{\text{m}^3}{\text{d}}\right)\left(\frac{1 \text{ d}}{24 \text{ h}}\right)}{5 \frac{\text{m}^3}{\text{m}^2 \cdot \text{h}}}$$
$$= 29.2 \text{ m}^2$$
$$\frac{A_s}{\text{filter}} = \frac{29.2 \text{ m}^2}{2}$$
$$= 14.6 \text{ m}^2$$

The gross production per filter is
$$\frac{3500 \frac{\text{m}^3}{\text{d}}}{2} = 1750 \text{ m}^3/\text{d}$$

The backwash losses per filter are
$$Q_B = v_B A_s t_B$$
$$= \left(30 \frac{\text{m}^3}{\text{m}^2 \cdot \text{h}}\right)(14.6 \text{ m}^2)\left(20 \frac{\text{min}}{\text{d}}\right)$$
$$\times \left(\frac{1 \text{ h}}{60 \text{ min}}\right)$$
$$= 146 \text{ m}^3/\text{d}$$

The conditioning losses per filter are
$$\left(5 \frac{\text{m}^3}{\text{m}^2 \cdot \text{h}}\right)(14.6 \text{ m}^2)\left(5 \frac{\text{min}}{\text{d}}\right)\left(\frac{1 \text{ h}}{60 \text{ min}}\right) = 6.1 \text{ m}^3/\text{d}$$

The net production per filter is
$$1750 \frac{\text{m}^3}{\text{d}} - 146 \frac{\text{m}^3}{\text{d}} - 6.1 \frac{\text{m}^3}{\text{d}} = 1598 \text{ m}^3/\text{d}$$
$$(1600 \text{ m}^3/\text{d})$$

The answer is (A).

127. Disregard sample 1 because DO_5 is below 2.0 mg/L (possible anaerobic activity), and disregard sample 4 because DO_5 is above 7.0 mg/L (insufficient aerobic activity).

For sample 2,
$$\frac{9.2 \frac{\text{mg}}{\text{L}} - 2.3 \frac{\text{mg}}{\text{L}}}{\frac{100 \text{ mL}}{300 \text{ mL}}} = 20.7 \text{ mg/L}$$

For sample 3,
$$\frac{9.3 \frac{\text{mg}}{\text{L}} - 5.8 \frac{\text{mg}}{\text{L}}}{\frac{50 \text{ mL}}{300 \text{ mL}}} = 21.0 \text{ mg/L}$$

BOD_5 at 25°C is
$$\frac{20.7 \frac{\text{mg}}{\text{L}} + 21.0 \frac{\text{mg}}{\text{L}}}{2} = 20.85 \text{ mg/L}$$
$$BOD_u = \frac{BOD_t}{(1 - e^{-kt})}$$
$$= \frac{20.85 \frac{\text{mg}}{\text{L}}}{1 - e^{-(0.4/\text{d})(5 \text{ d})}}$$
$$= 24.1 \text{ mg/L}$$
$$k_T = k\theta^{(T-25)}$$
$$k_{20} = \left(\frac{0.40}{\text{d}}\right)(1.047)^{20-25}$$
$$= 0.32/\text{d}$$

BOD_5 at 20°C is
$$\left(24.1 \frac{\text{mg}}{\text{L}}\right)\left(1 - e^{-(0.32/\text{d})(5\text{d})}\right) = 19.2 \text{ mg/L}$$
$$(19 \text{ mg/L})$$

The answer is (B).

128. Assume the river water is of good quality with low TDS and that DO is at saturation prior to the point of discharge.

Assume the river DO is at saturation just prior to receiving the discharge.

At 52°F, $DO_{\text{sat}} = 11.08$ mg/L.

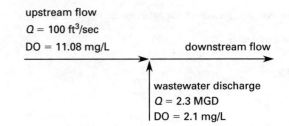

upstream flow
Q = 100 ft³/sec
DO = 11.08 mg/L

downstream flow

wastewater discharge
Q = 2.3 MGD
DO = 2.1 mg/L

$$\left(100 \, \frac{\text{ft}^3}{\text{sec}}\right)\left(11.08 \, \frac{\text{mg}}{\text{L}}\right)\left(28.25 \, \frac{\text{L}}{\text{ft}^3}\right)\left(86{,}400 \, \frac{\text{sec}}{\text{day}}\right)$$
$$+\left(2.3 \times 10^6 \, \frac{\text{gal}}{\text{day}}\right)\left(2.1 \, \frac{\text{mg}}{\text{L}}\right)\left(3.785 \, \frac{\text{L}}{\text{gal}}\right)$$
$$= \left(\begin{array}{c}\left(100 \, \frac{\text{ft}^3}{\text{sec}}\right)\left(28.25 \, \frac{\text{L}}{\text{ft}^3}\right)\left(86{,}400 \, \frac{\text{sec}}{\text{day}}\right) \\ + \left(2.3 \times 10^6 \, \frac{\text{gal}}{\text{day}}\right)\left(3.785 \, \frac{\text{L}}{\text{gal}}\right)\end{array}\right)$$
$$\times (\text{DO}_{\text{mix}})$$

$\text{DO}_{\text{mix}} = 10.8 \text{ mg/L}$

The answer is (C).

129. The travel time from the discharge point to a location 100 mi downstream is

$$\frac{(100 \text{ mi})\left(5280 \, \frac{\text{ft}}{\text{mi}}\right)}{\left(0.51 \, \frac{\text{ft}}{\text{sec}}\right)\left(86{,}400 \, \frac{\text{s}}{\text{d}}\right)} = 12 \text{ days}$$

$$\text{DO}_{\text{sat}} = 11.33 \text{ mg/L at } 50°\text{F}$$

At 12 days, the DO deficit is

$$11.33 \, \frac{\text{mg}}{\text{L}} - 9.1 \, \frac{\text{mg}}{\text{L}} = 2.23 \text{ mg/L} \quad (2.2 \text{ mg/L})$$

The answer is (A).

130.
$$\text{DO}_{\text{sat}} = 11.33 \text{ mg/L at } 50°\text{F}$$

From the DO sag curve, the critical DO deficit occurs at about 3.5 days and is equal to

$$\text{DO}_{\text{critical}} = \text{DO}_{\text{sat}} - \text{DO}_{\text{day},3.5}$$
$$= 11.33 \, \frac{\text{mg}}{\text{L}} - 6.1 \, \frac{\text{mg}}{\text{L}}$$
$$= 5.23 \text{ mg/L} \quad (5.2 \text{ mg/L})$$

The answer is (C).

131. From the DO sag curve, $t_c \approx 3.5$ days.

$$\frac{(3.5 \text{ days})\left(0.51 \, \frac{\text{ft}}{\text{sec}}\right)\left(86{,}400 \, \frac{\text{sec}}{\text{day}}\right)}{5280 \, \frac{\text{ft}}{\text{mi}}} = 29.2 \text{ mi}$$
$$(29 \text{ mi})$$

The answer is (C).

132. The annual TKN loading to the pond is

$$\frac{\left(2.23 \, \frac{\text{ft}^3}{\text{min}}\right)\left(1.43 \, \frac{\text{mg}}{\text{L}}\right)\left(28.25 \, \frac{\text{L}}{\text{ft}^3}\right)}{(24 \text{ ac})(8 \text{ ft})\left(\frac{1 \text{ yr}}{365 \text{ days}}\right)} \times \left(\frac{2.204 \text{ lbm}}{10^6 \text{ mg}}\right)\left(1440 \, \frac{\text{min}}{\text{day}}\right)$$

$$= 0.54 \text{ lbm/ac-ft yr}$$

The answer is (B).

133. Assume only organic nitrogen is lost to sediment. The organic nitrogen concentration in the inflow is

$$\text{organic N} = \text{TKN} - \text{ammonia N}$$
$$= 1.43 \, \frac{\text{mg}}{\text{L}} - 0.96 \, \frac{\text{mg}}{\text{L}}$$
$$= 0.47 \text{ mg/L}$$

Assume that inflow and outflow rates are equal at 2.23 ft^3/min.

From a mass balance of nitrogen,

$$(\text{inflow rate})\binom{\text{inflow TKN}}{\text{concentration}} - (\text{inflow rate})$$
$$\times \binom{\text{inflow organic nitrogen}}{\text{concentration}}$$
$$= (\text{outflow rate})\binom{\text{pond nitrogen}}{\text{concentration}}$$

$$\left(2.23 \, \frac{\text{ft}^3}{\text{min}}\right)\left(1.43 \, \frac{\text{mg}}{\text{L}}\right) - \left(2.23 \, \frac{\text{ft}^3}{\text{min}}\right)$$
$$\times \left(0.47 \, \frac{\text{mg}}{\text{L}}\right)$$
$$= \left(2.23 \, \frac{\text{ft}^3}{\text{min}}\right)\binom{\text{pond nitrogen}}{\text{concentration}}$$

$$\text{pond nitrogen concentration} = \frac{\left(2.23 \, \frac{\text{ft}^3}{\text{min}}\right)\left(1.43 \, \frac{\text{mg}}{\text{L}}\right) - \left(2.23 \, \frac{\text{ft}^3}{\text{min}}\right)\left(0.47 \, \frac{\text{mg}}{\text{L}}\right)}{2.23 \, \frac{\text{ft}^3}{\text{min}}}$$

$$= 1.43 \, \frac{\text{mg}}{\text{L}} - 0.47 \, \frac{\text{mg}}{\text{L}}$$
$$= 0.96 \text{ mg/L}$$

Note that if all the organic nitrogen is lost to sediment, the remaining nitrogen concentration in the pond is equal to the ammonia nitrogen concentration.

The answer is (C).

134. The mass of rubber supplied weekly is

$$\left(11\,000\,\frac{\text{kg}}{\text{d}}\right)\left(5\,\frac{\text{d}}{\text{wk}}\right) = 55\,000\,\text{kg/wk}$$

The mass of rubber supplied hourly to the incinerator is

$$\frac{55\,000\,\frac{\text{kg}}{\text{wk}}}{\left(7\,\frac{\text{d}}{\text{wk}}\right)\left(24\,\frac{\text{h}}{\text{d}}\right)} = 327\,\text{kg/h}$$

The hourly heat value of the shredded rubber is

$$\left(327\,\frac{\text{kg}}{\text{h}}\right)\left(90\,000\,\frac{\text{kJ}}{\text{kg}}\right) = 2.9 \times 10^7\,\text{kJ/h}$$

The answer is (B).

135. Assume that the sludge density is equal to the water density ($1000\,\text{kg/m}^3$).

$$\begin{aligned}
\text{sludge mass} &= (\text{volume})(\text{density}) \\
&\quad \times (1 - \text{fractional moisture}) \\
&= \left(4\,\frac{\text{m}^3}{\text{d}}\right)\left(1000\,\frac{\text{kg}}{\text{m}^3}\right)(1 - 0.76) \\
&\quad \times \left(\frac{1\,\text{d}}{24\,\text{h}}\right) \\
&= 40\,\text{kg/h}
\end{aligned}$$

The hourly heat value of the dry sludge solids is

$$\left(40\,\frac{\text{kg}}{\text{h}}\right)\left(30\,000\,\frac{\text{kJ}}{\text{kg}}\right) = 1.2 \times 10^6\,\text{kJ/h}$$

The answer is (B).

136.
$$c = \frac{q_b}{m\Delta T}$$

c	specific heat of water	kJ/kg °C
m	mass of water	kg
q_b	heat require to raise temperature to 100°C	kJ
q_v	heat required to vaporize water at its boiling point (heat of vaporization)	kJ/kg

$$c = 4.184\,\text{kJ/kg°C}$$
$$q_v = 2258\,\text{kJ/kg}$$

The temperature change is

$$\begin{aligned}
\Delta T &= 100°\text{C} - 20°\text{C} \\
&= 80°\text{C}
\end{aligned}$$

$$\begin{aligned}
m &= \left(4\,\frac{\text{m}^3}{\text{d}}\right)\left(1000\,\frac{\text{kg}}{\text{m}^3}\right)(0.76)\left(\frac{1\,\text{d}}{24\,\text{h}}\right) \\
&= 127\,\text{kg/h}
\end{aligned}$$

$$\begin{aligned}
q_b &= \left(4.184\,\frac{\text{kJ}}{\text{kg·°C}}\right)\left(127\,\frac{\text{kg}}{\text{h}}\right)(80°\text{C}) \\
&= 42\,509\,\text{kJ/h}
\end{aligned}$$

The total heat to dry the sludge is

$$\begin{aligned}
q_T &= q_b + q_v \\
&= 42\,509\,\frac{\text{kJ}}{\text{h}} + \left(2258\,\frac{\text{kJ}}{\text{kg}}\right)\left(127\,\frac{\text{kg}}{\text{h}}\right) \\
&= 329\,275\,\text{kJ/h} \quad (330\,000\,\text{kJ/h})
\end{aligned}$$

The answer is (D).

137. The total annual flow volume is

$$\begin{aligned}
&\left(0.7\,\frac{\text{m}^3}{\text{min}}\right)\left(60\,\frac{\text{min}}{\text{h}}\right)\left(16\,\frac{\text{h}}{\text{d}}\right)\left(20\,\frac{\text{d}}{\text{mo}}\right)\left(12\,\frac{\text{mo}}{\text{yr}}\right) \\
&= 161\,280\,\text{m}^3/\text{yr} \quad (160\,000\,\text{m}^3/\text{yr})
\end{aligned}$$

The answer is (A).

138.
$$\begin{aligned}
\text{Cr}_2(\text{SO}_4)_3\,\text{MW} &= (2)\left(52\,\frac{\text{g}}{\text{mol}}\right) + (3) \\
&\quad \times \left(32\,\frac{\text{g}}{\text{mol}} + (4)\left(16\,\frac{\text{g}}{\text{mol}}\right)\right) \\
&= 392\,\text{g/mol} \quad (392\,\text{mg/mmol})
\end{aligned}$$

$$\frac{406\,\frac{\text{mg}}{\text{L}}}{392\,\frac{\text{mg}}{\text{mmol}}} = 1.04\,\text{mmol/L}$$

From the precipitation reaction, 1.04 mmol/L of $\text{Cr}_2(\text{SO}_4)_3$ will react with (6)(1.04 mmol/L) of NaOH.

$$(6)\left(1.04\,\frac{\text{mmol}}{\text{L}}\right) = 6.24\,\text{mmol/L}$$

$$\begin{aligned}
\text{NaOH MW} &= 23\,\frac{\text{g}}{\text{mol}} + 16\,\frac{\text{g}}{\text{mol}} + 1\,\frac{\text{g}}{\text{mol}} \\
&= 40\,\text{g/mol} \quad (40\,\text{mg/mmol})
\end{aligned}$$

$$\frac{\left(40\ \dfrac{\text{mg}}{\text{mmol}}\right)\left(6.24\ \dfrac{\text{mmol}}{\text{L}}\right)\left(0.7\ \dfrac{\text{m}^3}{\text{min}}\right)\left(1000\ \dfrac{\text{L}}{\text{m}^3}\right)}{0.73}$$
$$\times \left(\dfrac{1\ \text{kg}}{10^6\ \text{mg}}\right)\left(1440\ \dfrac{\text{min}}{\text{d}}\right)\left(365\ \dfrac{\text{d}}{\text{yr}}\right)$$
$$= 125\,798\ \text{kg/yr}\quad(130\,000\ \text{kg/yr})$$

The answer is (D).

139. $\text{Cr}_2(\text{SO}_4)_3$ MW $= (2)\left(52\ \dfrac{\text{g}}{\text{mol}}\right) + (3)$
$$\times \left(32\ \dfrac{\text{g}}{\text{mol}} + (4)\left(16\ \dfrac{\text{g}}{\text{mol}}\right)\right)$$
$$= 392\ \text{g/mol}\quad(392\ \text{mg/mmol})$$

$$\dfrac{406\ \dfrac{\text{mg}}{\text{L}}}{392\ \dfrac{\text{mg}}{\text{mmol}}} = 1.04\ \text{mmol/L}$$

From the precipitation reaction, 1.04 mmol/L of $\text{Cr}_2(\text{SO}_4)_3$ will react to produce $(2)(1.04\ \text{mmol/L})$ of Cr(OH)_3.

$$(2)\left(1.04\ \dfrac{\text{mmol}}{\text{L}}\right) = 2.08\ \text{mmol/L}$$

Cr(OH)_3 MW $= 52\ \dfrac{\text{g}}{\text{mol}} + (3)\left(16\ \dfrac{\text{g}}{\text{mol}} + 1\ \dfrac{\text{g}}{\text{mol}}\right)$
$$= 103\ \text{g/mol}\quad(103\ \text{mg/mmol})$$

$$\left(103\ \dfrac{\text{mg}}{\text{mmol}}\right)\left(2.08\ \dfrac{\text{mmol}}{\text{L}}\right)\left(0.7\ \dfrac{\text{m}^3}{\text{min}}\right)$$
$$\times \left(1000\ \dfrac{\text{L}}{\text{m}^3}\right)\left(\dfrac{1\ \text{kg}}{10^6\ \text{mg}}\right)\left(1440\ \dfrac{\text{min}}{\text{d}}\right)$$
$$\times \left(365\ \dfrac{\text{d}}{\text{yr}}\right)$$
$$= 78\,823\ \text{kg/yr}\quad(79\,000\ \text{kg/yr})$$

The answer is (D).

140. The precipitation reaction for calcium hardness using NaOH as the reagent is

$$\text{Ca}^{+2} + 2\text{HCO}_3^- + 2\text{NaOH}$$
$$\rightarrow \text{CaCO}_3 + \text{Na}_2\text{CO}_3 + 2\text{H}_2\text{O}$$

$$\text{Ca}^{+2}\ \text{MW} = 40\ \text{mg/mmol}$$

$$\dfrac{187\ \dfrac{\text{mg}}{\text{L}}}{40\ \dfrac{\text{mg}}{\text{mmol}}} = 4.675\ \text{mmol/L}$$

From the reaction equation, 4.675 mmol/L Ca^{+2} reacts to with $(2)(4.675\ \text{mmol/L})$ NaOH.

$$(2)\left(4.675\ \dfrac{\text{mmol}}{\text{L}}\right) = 9.35\ \dfrac{\text{mmol}}{\text{L}}$$

NaOH MW $= 23\ \dfrac{\text{g}}{\text{mol}} + 16\ \dfrac{\text{g}}{\text{mol}} + 1\ \dfrac{\text{g}}{\text{mol}}$
$$= 40\ \text{g/mol}\quad(40\ \text{mg/mmol})$$

$$\left(40\ \dfrac{\text{mg}}{\text{mmol}}\right)\left(9.35\ \dfrac{\text{mmol}}{\text{L}}\right)\left(0.7\ \dfrac{\text{m}^3}{\text{min}}\right)$$
$$\times \left(1000\ \dfrac{\text{L}}{\text{m}^3}\right)\left(\dfrac{1\ \text{kg}}{10^6\ \text{mg}}\right)$$
$$\times \left(1440\ \dfrac{\text{min}}{\text{d}}\right)\left(365\ \dfrac{\text{d}}{\text{yr}}\right)$$
$$\overline{0.73}$$
$$= 188\,496\ \text{kg/yr}\quad(190\,000\ \text{kg/yr})$$

The answer is (D).

141. The precipitation reaction for calcium hardness using NaOH as the reagent is

$$\text{Ca}^{+2} + 2\text{HCO}_3^- + 2\text{NaOH}$$
$$\rightarrow \text{CaCO}_3 + \text{Na}_2\text{CO}_3 + 2\text{H}_2\text{O}$$

$$\text{Ca}^{+2}\ \text{MW} = 40\ \text{mg/mmol}$$

$$\dfrac{187\ \dfrac{\text{mg}}{\text{L}}}{40\ \dfrac{\text{mg}}{\text{mmol}}} = 4.675\ \text{mmol/L}$$

From the reaction equation, 4.675 mmol/L Ca^{+2} reacts to produce 4.675 mmol/L CaCO_3.

CaCO_3 MW $= 40\ \dfrac{\text{g}}{\text{mol}} + 12\ \dfrac{\text{g}}{\text{mol}} + (3)\left(16\ \dfrac{\text{g}}{\text{mol}}\right)$
$$= 100\ \text{g/mol}\quad(100\ \text{mg/mmol})$$

$$\left(100\ \dfrac{\text{mg}}{\text{mmol}}\right)\left(4.675\ \dfrac{\text{mmol}}{\text{L}}\right)\left(0.7\ \dfrac{\text{m}^3}{\text{min}}\right)\left(1000\ \dfrac{\text{L}}{\text{m}^3}\right)$$
$$\times \left(\dfrac{1\ \text{kg}}{10^6\ \text{mg}}\right)\left(1440\ \dfrac{\text{min}}{\text{d}}\right)\left(365\ \dfrac{\text{d}}{\text{yr}}\right)$$
$$= 172\,003\ \text{kg/yr}\quad(170\,000\ \text{kg/yr})$$

The answer is (D).

142. Assume that $\text{Cd}^{+2} + 2\text{OH}^- \rightarrow \text{Cd(OH)}_2$ represents cadmium hydroxide precipitation.

The initial concentration of Cd^{+2} is

$$\left(121\ \dfrac{\text{mg Cd}^{+2}}{\text{L}}\right)\left(\dfrac{1\ \text{mol}}{112.4\ \text{g}}\right)\left(\dfrac{1\ \text{g}}{1000\ \text{mg}}\right)$$
$$= 1.077 \times 10^{-3}\ \text{mol/L}$$

At pH 4.1,

$$[OH^-] = 10^{-(14-4.1)} \frac{mol}{L} = 1.26 \times 10^{-10} \text{ mol/L}$$

The desired equilibrium Cd^{+2} concentration is

$$\left(2.0 \frac{mg}{L}\right)\left(\frac{1 \text{ mol}}{112.4 \text{ g}}\right)\left(\frac{1 \text{ g}}{1000 \text{ mg}}\right)$$
$$= 1.78 \times 10^{-5} \text{ mol/L}$$

Assume a waste temperature of 25°C. At 25°C,

$$K_{sp} = [Cd^{+2}][OH^-]^2 = 5.27 \times 10^{-15}$$

The OH^- concentration at the desired equilibrium is

$$[OH^-]^2 = \frac{5.27 \times 10^{-15}}{1.78 \times 10^{-5} \frac{mol}{L}}$$
$$[OH^-] = 1.72 \times 10^{-5} \text{ mol/L}$$

The NaOH dose required to reduce the CD^{+2} concentration from 121 mg/L to 2 mg/L is

$$\left(1.72 \times 10^{-5} \frac{mol}{L} - 1.26 \times 10^{-10} \frac{mol}{L}\right)$$
$$\times \left(\frac{1 \text{ mol NaOH}}{1 \text{ mol OH}^-}\right)\left(40 \frac{g}{mol}\right)\left(1000 \frac{mg}{g}\right)$$
$$= 0.688 \text{ mg/L} \quad (0.69 \text{ mg/L})$$

The answer is (C).

143. At pH 4.1,

$$[OH^-] = 10^{-(14-4.1)} \frac{mol}{L} = 1.26 \times 10^{-10} \text{ mol/L}$$

At pH 8.6,

$$[OH^-] = 10^{-(14-8.6)} \frac{mol}{L} = 3.98 \times 10^{-6} \text{ mol/L}$$

The NaOH dose required to raise the pH to 8.6 is

$$\left(3.98 \times 10^{-6} \frac{mol}{L} - 1.26 \times 10^{-10} \frac{mol}{L}\right)$$
$$\times \left(\frac{1 \text{ mol NaOH}}{1 \text{ mol OH}^-}\right)\left(40 \frac{g}{mol}\right)\left(1000 \frac{mg}{g}\right)$$
$$= 0.159 \text{ mg/L} \quad (0.16 \text{ mg/L})$$

The answer is (B).

144. At pH 8.6,

$$[OH^-] = 10^{-(14-8.6)} \frac{mol}{L} = 3.98 \times 10^{-6} \text{ mol/L}$$

Assume a waste temperature of 25°C. At 25°C,

$$K_{sp} = [Cd^{+2}][OH^-]^2 = 5.27 \times 10^{-15}$$

The Cd^{+2} concentration at pH 8.6 is

$$[Cd^{+2}] = \frac{5.27 \times 10^{-15}}{\left(3.98 \times 10^{-6} \frac{mol}{L}\right)^2}$$
$$= 3.33 \times 10^{-4} \text{ mol/L}$$

$$\left(3.33 \times 10^{-4} \frac{mol}{L}\right)\left(112.4 \frac{g}{mol}\right)\left(1000 \frac{mg}{g}\right)$$
$$= 37.4 \text{ mg/L} \quad (37 \text{ mg/L})$$

The answer is (C).

145. The sum of the contaminant concentrations is

$$C_o = 1385 \text{ ppb} = 1.385 \text{ mg/L}$$

The chemical removed per day with C_e as the effluent concentration is

$$(C_o - C_e)Q = \left(1.385 \frac{mg}{L} - 0.1 \frac{mg}{L}\right)\left(92 \frac{gal}{min}\right)$$
$$\times \left(3.785 \frac{L}{gal}\right)\left(\frac{2.204 \text{ lbm}}{10^6 \text{ mg}}\right)$$
$$\times \left(1440 \frac{min}{day}\right)$$
$$= 1.42 \text{ lbm/day} \quad (1.4 \text{ lbm/day})$$

The answer is (C).

146. The sum of the contaminant concentrations is

$$C_o = 1385 \text{ ppb} = 1.385 \text{ mg/L}$$

At equilibrium in lead-follow mode operation, C_e is the influent concentration.

$$\frac{X}{M} = \frac{\text{mg chemical}}{\text{g GAC}}$$
$$= (2.837)\left(1.385 \frac{mg}{L}\right)^{0.431}$$
$$= \frac{3.26 \text{ mg chemical}}{\text{g GAC}} \quad (3.3 \text{ mg/g})$$

The answer is (D).

147. The GAC use rate is

$$\frac{\left(1.0 \; \frac{\text{lbm chemical}}{\text{day}}\right)\left(1000 \; \frac{\text{mg}}{\text{g}}\right)}{1.0 \; \frac{\text{mg chemical}}{\text{g GAC}}} = 1000 \; \text{lbm GAC/day}$$
$$(1000 \; \text{lbm/day})$$

The answer is (B).

148. The GAC change-out interval is

$$\frac{20{,}000 \; \frac{\text{lbm GAC}}{\text{vessel}}}{500 \; \frac{\text{lbm GAC}}{\text{day}}} = 40 \; \text{days/vessel}$$

The answer is (C).

149. The number of residences one truck can service in a single load is

$$\frac{\left(760 \; \frac{\text{lbm}}{\text{yd}^3}\right)(12 \; \text{yd}^3)}{\left(2.2 \; \frac{\text{lbm}}{\text{person-day}}\right)(7 \; \text{days})\left(4 \; \frac{\text{people}}{\text{residence}}\right)}$$
$$= 148 \; \text{residences/load}$$

The time required to collect a full load is

$$\left(148 \; \frac{\text{residences}}{\text{load}}\right)\left(\frac{32 \; \text{sec} + 7 \; \text{sec}}{1 \; \text{residence}}\right)\left(\frac{1 \; \text{min}}{60 \; \text{sec}}\right)$$
$$= 96 \; \text{min/load}$$

An 8 hr work day includes 480 min.

travel to route	38 min	38 min
collect load 1	96 min	134 min
travel to transfer station	38 min	172 min
unload	17 min	189 min
travel to route	38 min	227 min
collect load 2	96 min	323 min
travel to transfer station	38 min	361 min
unload	17 min	378 min

Insufficient time remains for the truck to collect a third load. One truck can collect two full loads in one day.

The answer is (B).

150. The total daily mass of waste generated by the community on an as-discarded basis is

$$(37{,}000 \; \text{people})\left(2.2 \; \frac{\text{lbm}}{\text{person-day}}\right) = 81{,}400 \; \text{lbm/day}$$

During a 5 day work week, the number of loads collected by a single truck is

$$\left(3 \; \frac{\text{loads}}{\text{truck-day}}\right)\left(5 \; \frac{\text{day}}{\text{wk}}\right) = 15 \; \text{loads/truck-wk}$$

$$\frac{\left(81{,}400 \; \frac{\text{lbm}}{\text{day}}\right)\left(7 \; \frac{\text{day}}{\text{wk}}\right)}{\left(760 \; \frac{\text{lbm}}{\text{yd}^3}\right)\left(12 \; \frac{\text{yd}^3}{\text{load}}\right)} = 62.48 \; \text{loads/wk}$$
$$(63 \; \text{loads/wk})$$

The required number of trucks is

$$\frac{63 \; \frac{\text{loads}}{\text{wk}}}{15 \; \frac{\text{loads}}{\text{truck-wk}}} = 4.2 \; \text{trucks} \quad (5 \; \text{trucks})$$

The answer is (B).

151. The graphical storage method is simple to use, compatible with SCS-NRCS runoff calculation methods, and best suited for small detention basin design, but it is less accurate than other methods.

The answer is (B).

152. According to 40 CFR 264.301(c)(1)(i)(B), the bottom liner layer must consist of 36 in of compacted clay.

Assuming the liner covers a flat surface, the required in-place volume of compacted soil is

$$\frac{(83 \; \text{ac})(36 \; \text{in})\left(43{,}560 \; \frac{\text{ft}^2}{\text{ac}}\right)}{\left(12 \; \frac{\text{in}}{\text{ft}}\right)\left(27 \; \frac{\text{ft}^3}{\text{yd}^3}\right)} = 401{,}720 \; \text{yd}^3$$
$$(400{,}000 \; \text{yd}^3)$$

The answer is (D).

153. The annual in-place waste volume landfilled is

$$\frac{\left(224{,}000 \; \frac{\text{lbm}}{\text{day}}\right)\left(365 \; \frac{\text{days}}{\text{yr}}\right)}{\left(1000 \; \frac{\text{lbm}}{\text{yd}^3}\right)} = 81{,}760 \; \text{yd}^3/\text{yr}$$

The annual soil cover volume is

$$\frac{81{,}760 \; \frac{\text{yd}^3}{\text{yr}}}{5} = 16{,}352 \; \text{yd}^3/\text{yr}$$

The total landfill volume at the 35 yr design capacity is the volume of waste plus the volume of soil cover.

$$\left(81{,}760 \; \frac{\text{yd}^3}{\text{yr}} + 16{,}352 \; \frac{\text{yd}^3}{\text{yr}}\right)(35 \; \text{yr})$$
$$= 3{,}433{,}920 \; \text{yd}^3 \quad (3{,}400{,}000 \; \text{yd}^3)$$

The answer is (D).

154. Waste 1 is hazardous because of toxicity. The regulatory concentration that, when exceeded, classifies a waste containing benzene as hazardous is 0.5 mg/L (40 CFR 261.24). Waste 1 contains benzene at 605 μg/L or 0.605 mg/L.

The answer is (D).

155. Waste 3 is hazardous because of corrosivity (pH less than or equal to 2.0) according to 40 CFR 261.22, and does not meet any other classification criteria. If the waste was neutralized to pH greater than 2.0, it would no longer exhibit a characteristic of hazardous waste.

The answer is (D).

156. The EPA has traditionally employed a command-and-control approach to implementing the CAA.

The answer is (A).

157. The sulfur feed to the plant is

$$\frac{\left(8.3 \frac{\text{lbm}}{\text{sec}}\right)(2.7\%)\left(86{,}400 \frac{\text{sec}}{\text{day}}\right)}{100\%} = 19{,}362 \text{ lbm/day}$$

The sulfur emitted to APC equipment is

$$\frac{\left(19{,}362 \frac{\text{lbm}}{\text{day}}\right)(100\% - 5\%)}{100\%} = 18{,}394 \frac{\text{lbm}}{\text{day}}$$

$$S + O_2 \rightarrow SO_2$$
$$S \text{ MW} = 32 \text{ g/mol}$$
$$SO_2 \text{ MW} = 32 \frac{\text{g}}{\text{mol}} + (2)\left(16 \frac{\text{g}}{\text{mol}}\right)$$
$$= 64 \text{ g/mol}$$

$$SO_2 \text{ emitted} = \frac{\left(18{,}394 \frac{\text{lbm}}{\text{d}} \text{ as S}\right)\left(64 \frac{\text{g}}{\text{mol}} \text{ as } SO_2\right)}{32 \frac{\text{g}}{\text{mol}} \text{ as S}}$$
$$= 36{,}788 \text{ lbm/day} \quad (37{,}000 \text{ lbm/day})$$

The answer is (D).

158. The average air flow for PM_{10} is the average of the initial and final air flows.

$$\frac{50 \frac{\text{ft}^3}{\text{min}} + 49.3 \frac{\text{ft}^3}{\text{min}}}{2} = 49.65 \text{ ft}^3/\text{min}$$

$$PM_{10} = \frac{(7.891 \text{ g} - 7.649 \text{ g})\left(10^6 \frac{\text{mg}}{\text{g}}\right)\left(35.29 \frac{\text{ft}^3}{\text{m}^3}\right)}{\left(49.65 \frac{\text{ft}^3}{\text{min}}\right)(24 \text{ hr})\left(60 \frac{\text{min}}{\text{hr}}\right)}$$
$$= 119 \text{ mg/m}^3 \quad (120 \text{ mg/m}^3)$$

The answer is (B).

159. The Ringelmann scale uses a percent opacity from completely transparent (Ringelmann no. 0) to completely opaque (Ringelmann no. 5). Each Ringelmann number corresponds to 20% opacity, so 5% opacity would represent a $^1/_4$ Ringelmann number. The Ringelmann number corresponding to 35% opacity is $1^3/_4$.

The answer is (B).

160. Most nitrogen oxide (NO) emissions occur from fuel release and fossil fuel combustion as relatively benign NO. However, NO can rapidly oxidize to nitrogen dioxide (NO_2) which is associated with respiratory ailments and which reacts with hydrocarbons in the presence of sunlight to form photochemical oxidants. NO_2 also reacts with the hydroxyl radical (•OH) in the atmosphere to form nitric acid (HNO_3), a contributor to acid rain.

When combusted, fossil fuels, mostly coal, release sulfur primarily as sulfur dioxide (SO_2) with much smaller amounts of sulfur trioxide (SO_3). Sulfur oxide aerosols may contribute to particulate matter concentrations in the air and are a constituent of acid rain. Sulfur oxides can significantly impact visibility. Sulfur dioxide also reacts with oxygen to form sulfur trioxide (SO_3), which subsequently reacts with water to form sulfuric acid (H_2SO_4). Sulfur dioxide may react directly with water in the atmosphere to form sulfurous acid (H_2SO_3).

The answer is (B).

161. Photochemical smog is produced from a variety of complex physical and chemical reactions involving nitrogen oxides, carbon monoxide, hydrocarbons, sunlight, and many other factors combined under favorable conditions. Among the photochemical oxidants involved in photochemical smog formation, ozone is the most common and is an irritant to the mucosa and lungs. Formaldehyde and acrolein are common aldehyde constituents of photochemical smog that are associated with eye irritation. Therefore, the primary nuisance components of photochemical smog include ozone and oxidized hydrocarbons, particularly aldehydes.

The answer is (C).

162. The efficiency is
$$E\% = (1 - e^{-k(-\Delta P)}) \times 100\%$$

$$k(-\Delta P) = \frac{3Z_o E_f \left(\dfrac{Q_l}{Q_g}\right)}{2d_d}$$

$$= \frac{(3)(4.25 \text{ m})(0.052)(0.6 \times 10^{-3})}{(2)(0.028 \text{ cm})\left(\dfrac{1 \text{ m}}{100 \text{ cm}}\right)}$$

$$= 0.71$$

$$E\% = (1 - e^{-0.71}) \times 100\%$$

$$= 51\%$$

The answer is (C).

163. The required cross-sectional area is

$$A_x = \frac{Q_g}{v_g} = \frac{10 \dfrac{\text{m}^3}{\text{s}}}{\left(34 \dfrac{\text{cm}}{\text{s}}\right)\left(\dfrac{1 \text{ m}}{100 \text{ cm}}\right)} = 29 \text{ m}^2$$

The answer is (B).

164. The pressure drops from the nozzle inlet value to zero as a free jet at the nozzle outlet.

$$-\Delta P = 0.05 \text{ atm} - 0 \text{ atm} = 0.05 \text{ atm}$$

The answer is (B).

165. Increasing the cross-sectional area will decrease the velocity of the gas, v_g, causing a decrease in efficiency.

The answer is (C).

166. The reaction between hydrogen cyanide and sodium hydroxide is

$$\text{HCN} + \text{NaOH} \rightarrow \text{NaCN} + \text{H}_2\text{O}$$

$$\text{HCN MW} = 1 \frac{\text{g}}{\text{mol}} + 12 \frac{\text{g}}{\text{mol}} + 14 \frac{\text{g}}{\text{mol}}$$

$$= 27 \text{ g/mol}$$

$$\left(1.94 \frac{\text{g}}{\text{m}^3}\right)\left(1.9 \frac{\text{m}^3}{\text{s}}\right) = 3.69 \text{ g/s}$$

$$\frac{3.69 \dfrac{\text{g}}{\text{s}}}{27 \dfrac{\text{g}}{\text{mol}}} = 0.137 \text{ mol/s}$$

0.137 mol/s of HCN will react with 0.137 mol/s NaOH. The sodium hydroxide flow rate is

$$\frac{\left(0.137 \dfrac{\text{mol}}{\text{s}}\right)\left(60 \dfrac{\text{s}}{\text{min}}\right)\left(1 \dfrac{\text{equiv}}{\text{mol}}\right)}{0.1 \dfrac{\text{equiv}}{\text{L}}} = 82 \text{ L/min}$$

The answer is (D).

167. The reaction between hydrogen cyanide and sodium hydroxide that produces sodium cyanide is

$$\text{HCN} + \text{NaOH} \rightarrow \text{NaCN} + \text{H}_2\text{O}$$

$$\text{HCN MW} = 1 \frac{\text{g}}{\text{mol}} + 12 \frac{\text{g}}{\text{mol}} + 14 \frac{\text{g}}{\text{mol}}$$

$$= 27 \text{ g/mol}$$

$$\left(1.94 \frac{\text{g}}{\text{m}^3}\right)\left(1.9 \frac{\text{m}^3}{\text{s}}\right) = 3.69 \text{ g/s}$$

$$\frac{3.69 \dfrac{\text{g}}{\text{s}}}{27 \dfrac{\text{g}}{\text{mol}}} = 0.137 \text{ mol/s}$$

0.137 mol/s HCN and NaOH will react to produce 0.137 mol/s NaCN.

$$\frac{\left(0.137 \dfrac{\text{mol}}{\text{s}}\right)\left(1 \dfrac{\text{equiv}}{\text{mol}}\right)}{\left(50 \dfrac{\text{L}}{\text{min}}\right)\left(\dfrac{1 \text{ min}}{60 \text{ s}}\right)} = 0.16 \text{ N}$$

The answer is (C).

168. The pressure drop is $-\Delta P$. The cleaning interval is $t = 5$ min.

$$-\Delta P = K_1 v_f + K_2 C_p v_f^2 t$$

$$= \frac{\left(\dfrac{0.32 \text{ kPa}}{1 \dfrac{\text{cm}}{\text{s}}}\right)\left(0.08 \dfrac{\text{m}^3}{\text{m}^2 \cdot \text{s}}\right)}{\dfrac{1 \text{ m}}{100 \text{ cm}}}$$

$$+ \left(\dfrac{0.98 \text{ kPa}}{\left(1 \dfrac{\text{cm}}{\text{s}}\right)\left(1 \dfrac{\text{g}}{\text{cm}^2}\right)}\right)$$

$$\times \frac{\left(21 \dfrac{\text{g}}{\text{m}^3}\right)\left(0.08 \dfrac{\text{m}^3}{\text{m}^2 \cdot \text{s}}\right)^2 (5 \text{ min})}{\left(100 \dfrac{\text{cm}}{\text{m}}\right)\left(\dfrac{1 \text{ min}}{60 \text{ sec}}\right)}$$

$$= 2.96 \text{ kPa}$$

Comparing 2.96 kPa to the desired range of 1.5 kPa to 2.0 kPa shows that a 5 min cleaning interval is too long.

The answer is (D).

169. The total fabric area is

$$A = \frac{Q_g}{v_f} = \frac{18 \dfrac{\text{m}^3}{\text{s}}}{0.08 \dfrac{\text{m}^3}{\text{m}^2 \cdot \text{s}}} = 225 \text{ m}^2$$

The area of each bag is

$$\pi(20 \text{ cm})\left(\frac{1 \text{ m}}{100 \text{ cm}}\right)(9 \text{ m}) = 5.65 \text{ m}^2$$

The number of bags is

$$\frac{225 \text{ m}^2}{5.65 \text{ m}^2} = 39.8 \quad (40)$$

The answer is (A).

170. Eight to ten compartments are typical for a total fabric area over the range from 1000 m² to 1600 m². Since 1500 m² is close to the upper boundary, choose ten compartments.

The answer is (B).

171. The figure shows an inversion to 500 m and subadiabatic conditions above 500 m.

The answer is (A).

172. The maximum mixing depth (MMD) extends from the ground surface to the elevation where the air temperature and plume temperature are equal. This corresponds to the intersection of the adiabatic lapse rate and the ambient lapse rate lines. The lapse rate intersection elevation is 1600 m, and the ground surface elevation is 100 m.

$$\text{MMD} = 1600 \text{ m} - 100 \text{ m}$$
$$= 1500 \text{ m}$$

The answer is (C).

173. The location of the maximum ground-level concentration of pollutants emitted from the stack will occur along the plume centerline.

From published reference tables, for a wind speed of 2.6 m/s measured 10 m above ground-level and with moderate incoming solar radiation, the stability category is B.

For emissions from a stack with an effective height H, the standard deviation along the vertical axis is

$$\sigma_z = 0.707H = (0.707)(120 \text{ m})$$
$$= 84.84 \quad (85)$$

Using published vertical dispersion coefficient plots for atmospheric stability category B and a standard deviation along the vertical axis of 85, the maximum ground-level concentration of pollutants emitted from the stack occurs at 800 m.

The answer is (B).

174. From published reference tables, for a wind speed of 2.6 m/s measured 10 m above ground level and with moderate incoming solar radiation, the stability category is B.

Using published vertical dispersion coefficient plots, for a 2000 m distance the standard deviation along the vertical axis is $\sigma_z = 230$ m.

Using published horizontal dispersion coefficient plots, for a 2000 m distance the standard deviation along the horizontal axis is $\sigma_y = 290$ m. The wind speed is $\mu = 2.6$ m/s, and emission rate is $Q_g = 5.4$ kg/s.

The plume centerline concentration at 2000 m downwind is

$$C_{2000,0} = \frac{Q_g e^{-0.5(H/\sigma_z)^2}}{\pi \mu \sigma_z \sigma_y}$$

$$= \frac{\left(5.4 \dfrac{\text{kg}}{\text{s}}\right)\left(10^6 \dfrac{\text{mg}}{\text{kg}}\right) \times \exp\left(-0.5\left(\dfrac{120 \text{ m}}{230 \text{ m}}\right)^2\right)}{\pi \left(2.6 \dfrac{\text{m}}{\text{s}}\right)(230 \text{ m})(290 \text{ m})}$$

$$= 8.7 \text{ mg/m}^3$$

The answer is (B).

175. From published atmospheric stability tables, for a wind speed of 2.6 m/s measured 10 m above ground level and moderate incoming solar radiation, the stability category is B.

Using published vertical dispersion coefficient plots, for a 1600 m distance the standard deviation along the vertical axis is $\sigma_z = 160$ m.

Using published horizontal dispersion coefficient plots, for a 1600 m distance the standard deviation along the horizontal axis is $\sigma_y = 240$ m.

The wind speed is $\mu = 2.6$ m/s, and the emission rate is $Q_g = 5.4$ kg/s.

The plume concentration at 1600 m downwind and 300 m cross-wind is

$$C_{1600,300,0} = \frac{Q_g \exp\left(-0.5\left(\frac{y}{\sigma_y}\right)^2\right) \times \exp\left(-0.5\left(\frac{H}{\sigma_z}\right)^2\right)}{\pi \mu \sigma_z \sigma_y}$$

$$= \frac{\left(5.4\,\frac{\text{kg}}{\text{s}}\right)\left(10^6\,\frac{\text{mg}}{\text{kg}}\right) \times \exp\left((-0.5)\left(\frac{300\text{ m}}{240\text{ m}}\right)^2\right) \times \exp\left((-0.5)\left(\frac{120\text{ m}}{160\text{ m}}\right)^2\right)}{\pi\left(2.6\,\frac{\text{m}}{\text{s}}\right)(160\text{ m})(240\text{ m})}$$

$$= 6.0\text{ mg/m}^3$$

The answer is (B).

176. For an inversion base, L, of 520 m and an 80 m stack height, H, the vertical dispersion coefficient is

$$\sigma_z = 0.47(L - H)$$
$$= (0.47)(520\text{ m} - 80\text{ m})$$
$$= 207\text{ m}$$

Using published vertical dispersion coefficient plots, for $\sigma_z = 207$ m and category D stability, the downwind distance is approximately 20 km.

The inversion form of the dispersion equation applies at approximately twice the distance at which the edge of the plume begins to interact with the inversion layer, or

$$(2)(20\text{ km}) = 40\text{ km}$$

The answer is (C).

177. The chronic daily intake is

$$\text{CDI} = \frac{C((\text{DI})(\%A))((\text{ED})(\%T))}{(\text{BW})(\text{LT})}$$

C	concentration	mg/L
DI	daily intake	L/d
ED	exposed duration	yr
BW	body weight	kg
LT	lifetime	yr
%A	percent absorbed	%
%T	percent of time exposed	%

Assume %A and %T are both 100%.

$$\text{CDI} = \frac{\left(0.270\,\frac{\text{mg}}{\text{L}}\right)\left(2\,\frac{\text{L}}{\text{d}} \times 100\%\right)(12\text{ yr} \times 100\%)}{(70\text{ kg})(70\text{ yr})}$$

$$= 1.3 \times 10^{-3}\,\frac{\text{mg}}{\text{kg}\cdot\text{d}}$$

$$\text{risk} = (\text{CDI})(\text{PF})$$
$$= \left(1.3 \times 10^{-3}\,\frac{\text{mg}}{\text{kg}\cdot\text{d}}\right)\left(1.75\left(\frac{\text{mg}}{\text{kg}\cdot\text{d}}\right)^{-1}\right)$$
$$= 2.3 \times 10^{-3}$$

The answer is (B).

178. The chronic daily intake is

$$\text{CDI} = \frac{C((\text{DI})(\%A))((\text{ED})(\%T))}{(\text{BW})(\text{LT})}$$

C	concentration	mg/m³
DI	daily intake	m³/d

$$\text{CDI} = \frac{\left(0.13 \times 10^{-3}\,\frac{\text{mg}}{\text{m}^3}\right)\left(20\,\frac{\text{m}^3}{\text{d}} \times 100\%\right) \times (12\text{ yr} \times 100\%)\left(30\,\frac{\text{min}}{\text{d}}\right)}{(70\text{ kg})(70\text{ yr})\left(1440\,\frac{\text{min}}{\text{d}}\right)}$$

$$= 1.3 \times 10^{-7}\text{ mg/kg}\cdot\text{d}$$

The risk is

$$\left(1.3 \times 10^{-7}\,\frac{\text{mg}}{\text{kg}\cdot\text{d}}\right)\left(50\left(\frac{\text{mg}}{\text{kg}\cdot\text{d}}\right)^{-1}\right) = 6.6 \times 10^{-6}$$

The answer is (B).

179. The hazard quotient is the average daily dose during exposure period divided by the reference dose.

$$\frac{\left(0.270\,\frac{\text{mg}}{\text{L}}\right)\left(1\,\frac{\text{L}}{\text{d}}\right)}{(15\text{ kg})\left(0.0003\,\frac{\text{mg}}{\text{kg}\cdot\text{d}}\right)} = 60$$

The answer is (C).

180. The arsenic concentration in fish is

$$\left(\begin{array}{c}\text{concentration}\\\text{in water}\end{array}\right)(\text{BCF}) = \left(0.270\ \frac{\text{mg}}{\text{L}}\right)\left(44\ \frac{\text{L}}{\text{kg}}\right)$$
$$= 12\ \text{mg arsenic/kg fish}$$

$$\text{CDI} = \frac{C((\text{DI})(\%A))((\text{ED})(\%T))}{(\text{BW})(\text{LT})}$$

C	concentration	mg/kg
DI	daily intake	g/d

$$\text{CDI} = \frac{\left(12\ \frac{\text{mg}}{\text{kg}}\right)\left(54\ \frac{\text{g}}{\text{d}} \times 100\%\right) \times (12\ \text{yr} \times 100\%)\left(\frac{1\ \text{kg}}{1000\ \text{g}}\right)}{(70\ \text{kg})(70\ \text{yr})}$$
$$= 1.6 \times 10^{-3}\ \text{mg/kg}\cdot\text{d}$$

$$\text{risk} = (\text{CDI})(\text{PF})$$
$$= \left(1.6 \times 10^{-3}\ \frac{\text{mg}}{\text{kg}\cdot\text{d}}\right)\left(1.75\left(\frac{\text{mg}}{\text{kg}\cdot\text{d}}\right)^{-1}\right)$$
$$= 2.8 \times 10^{-3}$$

The answer is (B).

181. The cumulative cancer risk from all exposures is

$$9.2 \times 10^{-5} + 1.7 \times 10^{-7} + 1.8 \times 10^{-4} = 2.7 \times 10^{-4}$$

The answer is (C).

182. For purposes of the MSDS relevant to the workplace, hazardous materials are those listed in 29 CFR 1910, Subpart Z, also known as the Z list.

The answer is (C).

183. The MSDS identifies workplace ventilation, personal breathing apparatuses, and personal protective clothing requirements.

The answer is (D).

184. The heading of the MSDS provides the name, address, and contact information for the material manufacturer or supplier, the name of the material covered by the MSDS, and the date that the MSDS was issued or revised.

The answer is (D).

185. The material name on the MSDS must always match the name printed on the material container.

The answer is (D).

186. The MSDS includes a complete list of all hazardous materials included in the mixture, regardless of their concentration or carcinogenicity. Inert materials (those that pose no hazard) that may be in the mixture do not need to be specifically listed.

The answer is (B).

187. Protective equipment for preventing employee exposure to air contaminants is an acceptable control measure only when administrative and engineering controls cannot be implemented.

The answer is (B).

188. An employee may only be exposed to an air contaminant listed with an acceptable ceiling concentration and maximum peak concentration and duration when the exposure does not exceed the maximum peak concentration and the exposure is less than the listed duration.

The answer is (D).

189. The average velocity of the groundwater is

$$v_{aw} = \frac{Ki}{n_e} = \frac{\left(123\ \frac{\text{ft}}{\text{day}}\right)\left(0.00091\ \frac{\text{ft}}{\text{ft}}\right)}{(0.31)\left(3.28\ \frac{\text{ft}}{\text{m}}\right)}$$
$$= 0.11\ \text{m/d}$$

The answer is (B).

190.
$$K_d = f_{oc}K_{oc}$$
$$= \left(198\ \frac{\text{mg}}{\text{kg}}\right)\left(10^{-6}\ \frac{\text{kg}}{\text{mg}}\right)\left(91\ \frac{\text{mL}}{\text{g}}\right)$$
$$\times \left(1\ \frac{\text{cm}^3}{\text{mL}}\right)$$
$$= 0.018\ \text{cm}^3/\text{g}$$

$$r_f = 1 + \frac{B_d K_d}{n_e}$$
$$= 1 + \frac{\left(1.65\ \frac{\text{g}}{\text{cm}^3}\right)\left(0.018\ \frac{\text{cm}^3}{\text{g}}\right)}{0.31}$$
$$= 1.096$$

The average velocity of benzene relative to the groundwater velocity is

$$v_{ab} = \frac{v_{aw}}{r_f} = \frac{v_{aw}}{1.096}$$
$$= 0.91 v_{aw}$$

The answer is (D).

191. The following equation applies when the flow path, L, is less than 3500 m. The longitudinal dynamic dispersity is

$$\alpha_L = 0.0175 L^{1.46}$$
$$= (0.0175)(100 \text{ m})^{1.46}$$
$$= 14.6 \quad (15)$$

The answer is (D).

192. The longitudinal hydrodynamic dispersion is

$$D_L = \alpha_L v_{ab} + D^*$$

D^* is the effective diffusion coefficient and can be neglected in this case since the soil is permeable ($K = 123$ ft/day).

$$D_L = (1.0 \text{ m})\left(1.0 \frac{\text{m}}{\text{d}}\right) = 1.0 \text{ m}^2/\text{d}$$

The answer is (C).

193.
$$C = \frac{C_o}{2} \text{erfc}\left(\frac{L - v_{ab}t}{2\sqrt{D_L t}}\right)$$

The erfc is the complementary error function.

$$\text{erfc}\left(\frac{L - v_{ab}t}{2\sqrt{D_L t}}\right) = \text{erfc}\left(\frac{100 \text{ m} - \left(1.0 \frac{\text{m}}{\text{d}}\right)(90 \text{ d})}{(2)\sqrt{\left(1.0 \frac{\text{m}^2}{\text{d}}\right)(90 \text{ d})}}\right)$$
$$= \text{erfc } 0.527$$
$$= 0.45$$

The benzene concentration is

$$C = \frac{\left(1000 \frac{\mu\text{g}}{\text{L}}\right)(0.45)}{2}$$
$$= 225 \ \mu\text{g/L} \quad (230 \ \mu\text{g/L})$$

The answer is (C).

194. Vinyl chloride is an ethylene with a condensed structural formula of CH_2CHCl. It follows the sequence of ethylenes from perchloroethylene (CCl_2CCl_2) to trichloroethylene ($CHClCCl_2$) to dichloroethylene ($CHClCHCl$). The compounds in both sequences containing chloroform are methanes, and the chemicals in the sequence starting with tetrachloroethane are ethanes.

The answer is (C).

195. The human olfactory sense can detect some common compounds associated with municipal sewage at concentrations below 0.0005 ppm and can recognize some odors at concentrations near 0.0005 ppm.

The answer is (A).

196. The primary method employed to characterize and measure odors is organoleptic or sensory detection methods.

The answer is (B).

197.
$$\text{TON} = \frac{A + B}{A}$$

A is the volume of sample in milliliters and B is the volume of odor-free dilution water in milliliters.

Since $A = 25$ mL and $B = 200$ mL $- 25$ mL $= 175$ mL,

$$\text{TON} = \frac{25 \text{ mL} + 175 \text{ mL}}{25 \text{ mL}} = 8$$

The answer is (C).

198. Radon mitigation measures are directed at foundations, slab flooring placed on grade, and crawl spaces where close contact with radon-containing soils is most likely. Attics are typically not a target of radon abatement.

The answer is (C).

199. Radon may be detected by certified professionals, by ordinary citizens using over-the-counter test kits, or by permanently installed monitoring devices.

The answer is (D).

200. Radon exposure in homes presents the greatest risk because the largest percentage of people's time is generally spent in the home.

The answer is (B).

Instructions

Name: _____
      Last   First   Middle Initial

Do not enter solutions in the test booklet. Complete solutions must be entered on the answer sheet provided by your proctor.

This is an open-book examination. You may use textbooks, handbooks, and other bound references, along with a battery-operated, silent, nonprinting calculator. Unbound reference materials and notes, scratchpaper, and writing tablets are not permitted. You may not consult with or otherwise share any materials or information with others taking the exam.

You must work all 50 multiple-choice problems in the four-hour period allocated for the morning session. Each of the 50 problems is worth one point. No partial credit will be awarded. Your score will be based entirely on the responses marked on the answer sheet. You may use blank spaces in the exam booklet for scratch work. However, no credit will be awarded for work shown in margins or on other pages of the exam booklet. Mark only one answer to each problem.

Principles and Practice of Engineering Examination

MORNING SESSION
Sample Examination 3

201.	(A)	(B)	(C)	(D)	226.	(A)	(B)	(C)	(D)
202.	(A)	(B)	(C)	(D)	227.	(A)	(B)	(C)	(D)
203.	(A)	(B)	(C)	(D)	228.	(A)	(B)	(C)	(D)
204.	(A)	(B)	(C)	(D)	229.	(A)	(B)	(C)	(D)
205.	(A)	(B)	(C)	(D)	230.	(A)	(B)	(C)	(D)
206.	(A)	(B)	(C)	(D)	231.	(A)	(B)	(C)	(D)
207.	(A)	(B)	(C)	(D)	232.	(A)	(B)	(C)	(D)
208.	(A)	(B)	(C)	(D)	233.	(A)	(B)	(C)	(D)
209.	(A)	(B)	(C)	(D)	234.	(A)	(B)	(C)	(D)
210.	(A)	(B)	(C)	(D)	235.	(A)	(B)	(C)	(D)
211.	(A)	(B)	(C)	(D)	236.	(A)	(B)	(C)	(D)
212.	(A)	(B)	(C)	(D)	237.	(A)	(B)	(C)	(D)
213.	(A)	(B)	(C)	(D)	238.	(A)	(B)	(C)	(D)
214.	(A)	(B)	(C)	(D)	239.	(A)	(B)	(C)	(D)
215.	(A)	(B)	(C)	(D)	240.	(A)	(B)	(C)	(D)
216.	(A)	(B)	(C)	(D)	241.	(A)	(B)	(C)	(D)
217.	(A)	(B)	(C)	(D)	242.	(A)	(B)	(C)	(D)
218.	(A)	(B)	(C)	(D)	243.	(A)	(B)	(C)	(D)
219.	(A)	(B)	(C)	(D)	244.	(A)	(B)	(C)	(D)
220.	(A)	(B)	(C)	(D)	245.	(A)	(B)	(C)	(D)
221.	(A)	(B)	(C)	(D)	246.	(A)	(B)	(C)	(D)
222.	(A)	(B)	(C)	(D)	247.	(A)	(B)	(C)	(D)
223.	(A)	(B)	(C)	(D)	248.	(A)	(B)	(C)	(D)
224.	(A)	(B)	(C)	(D)	249.	(A)	(B)	(C)	(D)
225.	(A)	(B)	(C)	(D)	250.	(A)	(B)	(C)	(D)

Exam 3—Morning Session

SITUATION FOR PROBLEMS 201–203

A drainage basin has the following characteristics.

total area that will discharge runoff to the detention basin	100 ha
total area of the developed site	
woodlands	40 ha
rooftops, roads, driveways, etc.	20 ha
landscaping	40 ha
runoff coefficients	
pre-developed site (woodlands)	0.25
rooftops, roads, driveways, etc.	0.95
landscaping (average slope)	0.15
25 yr rainfall intensity	9.1 cm/h
2 yr rainfall intensity	5.3 cm/h

201. What is the pre-development peak discharge at the 2 yr storm? Use the rational equation method.

(A) 0.5 m^3/s
(B) 2.2 m^3/s
(C) 3.7 m^3/s
(D) 6.3 m^3/s

202. What is the post-development peak discharge of the 25 yr storm?

(A) 5.2 m^3/s
(B) 8.8 m^3/s
(C) 11 m^3/s
(D) 13 m^3/s

203. Does the developed site likely require storm water control and containment devices if the pre-development flow is 1 m^3/s and the post-development flow is 1.5 m^3/s?

(A) yes, because control and containment are always provided when post-development exceeds pre-development flow
(B) yes, because post-development flow is considerably greater than pre-development flow
(C) no, because post-development flow is less than twice pre-development flow
(D) no, because post-development and pre-development flows are very small

SITUATION FOR PROBLEMS 204–205

Bench scale bioreactor tests have been conducted using samples of a domestic wastewater in anticipation of complete mix activated sludge process design. The tests, conducted at 20°C without solids recycle, produced the following results for biochemical oxygen demand (BOD), hydraulic residence time, and volatile suspended solids (VSS).

test reactor	influent soluble BOD$_5$ (mg/L)	effluent soluble BOD$_5$ (mg/L)	hydraulic residence time (days)	reactor VSS (mg/L)
1	275	8	3.5	133
2	275	15	1.9	131
3	275	21	1.6	135
4	275	32	1.3	130
5	275	44	1.0	124

204. What is the value of the half-velocity constant at 20°C?

(A) 0.29 mg/L
(B) 3.4 mg/L
(C) 12 mg/L
(D) 40 mg/L

205. What is the value of the maximum yield coefficient?

(A) 0.035
(B) 0.044
(C) 0.55
(D) 1.9

SITUATION FOR PROBLEMS 206–209

A beef slaughterhouse is encountering problems operating their existing wastewater treatment pond system. The slaughterhouse is unable to satisfy the permit requirements for discharge of the system effluent to the local municipal wastewater treatment plant. Effluent quality required for the discharge is 200:200 for the BOD-to-total-suspended-solids (TSS) ratio (all units in mg/L). The current wastewater characteristics generated by the slaughterhouse are

parameter	concentration (mg/L)
BOD$_5$	2000
TSS	1800

The wastewater flow is 400 m^3/d, and essentially all discharges to the pond system occur during normal operating hours (an 8 h day).

The pond system consists of three ponds operating in series. Pond 1 is an anaerobic pond based on a volumetric mass loading of 0.35 kg BOD/m^3·d. Ponds 2 and 3 are intended to operate as aerobic ponds; however, they are not mechanically aerated. A schematic of the pond system follows.

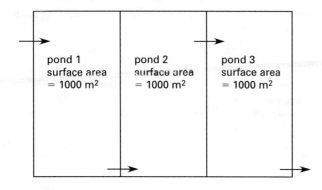

206. What minimum liquid depth should be maintained in pond 1 to satisfy volumetric mass loading criteria?

(A) 1.5 m
(B) 2.3 m
(C) 3.0 m
(D) 7.6 m

207. How could the BOD removal efficiency of pond 1 be improved?

(A) through pretreatment to remove scum-forming matter
(B) by aerating ponds 2 and 3
(C) by equalizing influent flow to the pond system
(D) by installing mixers to mix pond 1 contents

208. Assuming a 50% BOD removal efficiency in pond 1 and a 1-to-1 ratio of dissolved oxygen required for BOD removal, what would be the minimum theoretical aeration requirement for an additional 80% BOD removal through pond 2?

(A) 13 kg DO/h
(B) 16 kg DO/h
(C) 27 kg DO/h
(D) 33 kg DO/h

209. What would be the consequence of installing grease removal equipment ahead of the discharge to pond 1?

(A) It would improve BOD removal by reducing scum formation.
(B) It would reduce soluble BOD loading.
(C) It would not materially influence pond performance.
(D) It would decrease BOD removal by reducing scum formation.

SITUATION FOR PROBLEMS 210–211

An industrial wastewater effluent is treated using a complete mix activated sludge process followed by a clarifier. The clarifier produces a VSS discharge of 30 mg/L. The total flow rate to the bioreactor is 11 MGD and contains 160 mg/L VSS. The bioreactor volume is 5800 m^3 and is operated to maintain a mixed liquor volatile suspended solids (MLVSS) of 2100 mg/L with biomass production at 0.45 kg/m^3·d. The return volatile solids concentration is 14 000 mg/L. The specific gravity of the volatile solids is 1.0.

210. What is the total return solids flow rate?

(A) 0.15 MGD
(B) 0.18 MGD
(C) 1.7 MGD
(D) 5.0 MGD

211. What is the total wasted sludge flow rate when the return sludge flow rate is 1.0 MGD?

(A) 0.18 MGD
(B) 0.63 MGD
(C) 0.77 MGD
(D) 1.4 MGD

SITUATION FOR PROBLEMS 212–213

A wastewater is treated using a complete mix activated sludge process followed by clarification. NPDES permit requirements limit discharges of VSS to 30 mg/L and total phosphorous to 1.0 mg/L. The plant routinely violates its discharge limits with VSS averaging 60 mg/L and total phosphorous averaging 1.3 mg/L. The source of the violations is believed to be associated with the operation of the two secondary clarifiers.

212. Which of the following actions would not result in improved performance of the secondary clarifiers?

(A) constructing a third secondary clarifier
(B) feeding chemicals to improve settling of the biofloc
(C) decreasing the depth of the inlet baffles
(D) increasing the length of the effluent weirs

213. If effluent TSS from the secondary clarifiers is reduced from 60 mg/L to 30 mg/L, what will happen to the total phosphorous concentration in the effluent?

(A) It will not change.
(B) It will decrease to less than 1.0 mg/L.
(C) It will decrease by half to 0.65 mg/L.
(D) It will decrease by an indeterminate amount.

SITUATION FOR PROBLEMS 214–218

Bench tests to evaluate the mass transfer coefficient using an appropriate packing material produced the following results.

time (s)	concentration (μg/L)
0	2000
20	1620
60	1023
120	545
240	134
360	45
420	18

The chemical characteristics are as follows.
molecular weight   120 g/mol
vapor pressure   0.26 atm at 25°C
solubility in water 9300 mg/L at 25°C

Assume a first order reaction stripping factor of 3.0, an air temperature of 25°C, and an atmospheric pressure of 1.0 atm.

214. What is the value of the mass transfer coefficient?

(A) 0.011 s^{-1}
(B) 0.21 s·L/μg
(C) 4.7 μg/L·s
(D) 18 (unitless)

215. What is the value of Henry's constant in unitless form?

(A) 0.00050
(B) 0.00335
(C) 0.14
(D) 0.21

216. What is the air-to-water ratio when Henry's constant is 0.05?

(A) 14
(B) 22
(C) 60
(D) 89

217. To which parameter is air stripper design most sensitive for a given chemical?

(A) stripping factor
(B) mass transfer coefficient
(C) Henry's constant
(D) air-to-water ratio

218. What would be the impact of recirculating flow?

(A) Henry's constant would increase.
(B) The air to water ratio would increase.
(C) The mass transfer coefficient would increase.
(D) The hydraulic loading rate (HLR) would increase.

SITUATION FOR PROBLEMS 219–220

Water samples characterized by the following analysis were collected from a well intended to provide 500 gal/min flow to a manufacturing facility.

ion	concentration (mg/L)
SO_4^{-2}	28
Ca^{+2}	67
Mg^{+2}	21
HCO_3^-	353

219. What is the daily requirement of NaOH (100% purity) for softening the water?

(A) 440 kg/d
(B) 560 kg/d
(C) 1000 kg/d
(D) 1700 kg/d

220. What is the daily sludge volume at 35% solids generated from softening the water?

(A) 0.16 m^3/d
(B) 1.7 m^3/d
(C) 16 m^3/d
(D) 46 m^3/d

SITUATION FOR PROBLEMS 221–225

A sedimentation basin is required to treat 10 000 m^3/d of flow containing 234 mg/L TSS. The settling zone depth is 2.5 m, the settling time is 100 min, and the settling zone length-to-width ratio is 3:1. Settling characteristics are as described in the following illustration.

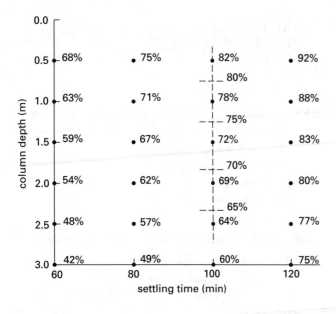

221. What would be the dimensions of the sedimentation basins if inlet and outlet zones of 3 m and a sludge zone of 0.5 m were included?

(A) $l = 27$ m, $w = 9$ m, $d = 2.5$ m
(B) $l = 33$ m, $w = 11$ m, $d = 3.0$ m
(C) $l = 35$ m, $w = 9.6$ m, $d = 3.0$ m
(D) $l = 39$ m, $w = 11$ m, $d = 3.0$ m

222. What is the average particle horizontal settling velocity for a settling zone length of 30 m?

(A) 4.7 m/h
(B) 5.6 m/h
(C) 6.9 m/h
(D) 14 m/h

223. What is the effect on efficiency if the settling zone depth is decreased to 2.0 m?

(A) decreases
(B) increases
(C) remains unchanged if settling time remains unchanged
(D) remains unchanged if overflow rate remains unchanged

224. What is the effect on efficiency if the settling time is increased to 120 min?

(A) decreases
(B) increases
(C) remains unchanged if settling depth remains unchanged
(D) remains unchanged is overflow rate remains unchanged

225. Assuming 80% efficiency and 30% solids, what is the daily volume of sludge requiring disposal?

(A) 0.56 m^3/d
(B) 1.9 m^3/d
(C) 6.2 m^3/d
(D) 7.8 m^3/d

SITUATION FOR PROBLEM 226

The average daily water use characteristics of a community are presented in the following illustration.

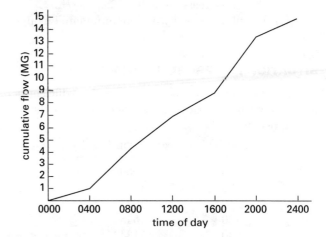

226. What is the daily demand?

(A) 0.63 MGD
(B) 2.5 MGD
(C) 11 MGD
(D) 15 MGD

SITUATION FOR PROBLEMS 227–228

A 200 mL water sample was titrated with 0.03N H$_2$SO$_4$. The initial pH was 7.8, and 28 mL of acid was added to reach pH 4.5.

227. What is the total alkalinity of the sample expressed in mg/L as CaCO$_3$?

(A) 110 mg/L as CaCO$_3$
(B) 140 mg/L as CaCO$_3$
(C) 210 mg/L as CaCO$_3$
(D) 420 mg/L as CaCO$_3$

228. What are the dominant alkalinity species present in the sample?

(A) carbonate only
(B) bicarbonate only
(C) bicarbonate and carbonate
(D) bicarbonate, carbonate, and hydroxide

SITUATION FOR PROBLEMS 229–233

Water samples were collected and submitted to a commercial analytical laboratory for analysis. The laboratory provided the following results.

ion	concentration (mg/L)
SO_4^{-2}	17
Na^+	48
K^+	14
Cl^-	39
Ca^{+2}	56
Mg^{+2}	12
HCO_3^-	237

total dissolved solids (TDS)	425 mg/L
specific conductivity	32 μS

229. Is the ion analysis complete?

(A) yes
(B) no, it is deficient in anions
(C) no, it is deficient in cations
(D) cannot be determined

230. What is the total hardness concentration?

(A) 50 mg/L as $CaCO_3$
(B) 140 mg/L as $CaCO_3$
(C) 190 mg/L as $CaCO_3$
(D) 210 mg/L as $CaCO_3$

231. What is the carbonate hardness concentration when the total hardness is 100 mg/L as $CaCO_3$?

(A) 50 mg/L as $CaCO_3$
(B) 74 mg/L as $CaCO_3$
(C) 100 mg/L as $CaCO_3$
(D) 190 mg/L as $CaCO_3$

232. What is the noncarbonate hardness concentration when both the total hardness and the carbonate hardness are 100 mg/L as $CaCO_3$?

(A) 0 mg/L as $CaCO_3$
(B) 50 mg/L as $CaCO_3$
(C) 100 mg/L as $CaCO_3$
(D) 200 mg/L as $CaCO_3$

233. What materials comprise the majority of the TDS of the sample?

(A) ionic solids
(B) nonionic solids
(C) approximately equal amounts of both ionic and nonionic solids
(D) cannot be determined

SITUATION FOR PROBLEMS 234–235

A community generates solid waste with the following characteristics.

component	mass (%)	moisture (%)	discarded density (kg/m^3)	discarded energy (kJ/kg)	ash (%)
food	16	70	290	4650	5
glass	6	2	195	150	98
plastic	5	2	65	32 600	10
paper	44	6	85	16 750	6
ferrous metal	5	3	320	840	98
nonferrous metal	5	2	160	700	96
yard clippings	19	60	105	6500	4.5

234. What is the energy content of the discarded bulk waste?

(A) 1700 kJ/kg
(B) 9300 kJ/kg
(C) 11 000 kJ/kg
(D) 13 000 kJ/kg

235. What is the energy content of the dry bulk waste?

(A) 1700 kJ/kg
(B) 9300 kJ/kg
(C) 11 000 kJ/kg
(D) 13 000 kJ/kg

SITUATION FOR PROBLEMS 236–239

Granular activated carbon (GAC) was selected to remove an organic solvent from contaminated groundwater. It is expected that an extraction well system will be able to produce about 500 m^3/d of continuous flow and will result in a GAC use rate of 280 kg/d. Two adsorber vessels of 10 000 kg GAC capacity each are used in a lead-follow configuration. The vessel bed volume is 20 m^3, and the bed diameter is 3 m.

New GAC can be purchased for $3.00/kg, and reactivation of spent carbon is available for $2.50/kg. Carbon costs include transportation. New carbon make-up per reactivation cycle is 15%.

236. What is the adsorption vessel empty bed contact time (EBCT)?

(A) 29 min
(B) 58 min
(C) 87 min
(D) 120 min

237. What is the adsorption vessel hydraulic loading rate (HLR)?

(A) 35 m^3/m^2·d
(B) 71 m^3/m^2·d
(C) 110 m^3/m^2·d
(D) 140 m^3/m^2·d

238. What is the first year GAC cost, including initial fill, assuming continuous operation?

(A) $220,000
(B) $260,000
(C) $320,000
(D) $360,000

239. How would GAC use be impacted if flow recirculation was included?

(A) no change, since the mass loading would not change
(B) increase, since the hydraulic loading rate would increase
(C) increase, since the influent concentration would decrease
(D) decrease, since the influent concentration would decrease

SITUATION FOR PROBLEMS 240-241

A municipality operates an anaerobic digester and it recovers the methane gas to offset energy costs. To increase the value of the digester gases, the city is considering separating carbon dioxide from other digester gases and then selling it. The digester produces about 60,000 ft^3 of gas daily at 1 atm and 25°C, with about 25% of the gas being carbon dioxide.

240. What is the daily mass of carbon dioxide generated by the digester?

(A) 760 kg/d
(B) 830 kg/d
(C) 3300 kg/d
(D) 4200 kg/d

241. How many 1 L bottles of beverage per 100 kg of recovered carbon dioxide can be carbonated at 10°C and 1.2 atm?

(A) 3000 L/100 kg
(B) 6100 L/100 kg
(C) 25 000 L/100 kg
(D) 35 000 L/100 kg

SITUATION FOR PROBLEMS 242-244

A metal plating process produces a wastewater with the following characteristics.

flow	1000 m^3/d
influent temperature	43°C
lead (Pb(II)) concentration	72 mg/L

242. Will sodium hydroxide precipitate Pb(II)?

(A) Yes, the standard free energy change, $\Delta G°$, for the reaction with Pb(II) as a reactant is positive.
(B) Yes, the standard free energy change, $\Delta G°$, for the reaction with Pb(II) as a reactant is negative.
(C) No, the standard free energy change, $\Delta G°$, for the reaction with Pb(II) as a reactant is positive.
(D) No, the standard free energy change, $\Delta G°$, for the reaction with Pb(II) as a reactant is negative.

243. What is the solubility product for lead hydroxide at 43°C?

(A) 7.2×10^{-17}
(B) 2.5×10^{-16}
(C) 3.7×10^{16}
(D) 2.8×10^{35}

244. What is the dominant Pb(II) specie at pH 7.6?

(A) Pb^{+2}
(B) $PbOH^+$
(C) $Pb(OH)_2$
(D) $Pb(OH)_3^-$

SITUATION FOR PROBLEM 245

The results of a time study and route analysis for curbside residential waste collection are as follows.

population	1400
solid waste generation rate	3.6 lbm/day·person
number of residences	388
average driving time between residences	18 sec
average pick-up/load time at each residence	32 sec
travel time from truck yard to route start	28 min
average travel time between route and landfill	47 min
time to unload at landfill	15 min
travel time from landfill to truck yard	34 min
truck compacted waste capacity	10 yd^3
truck compaction ratio	3.1
typical waste as-discarded density	240 lbm/yd^3

245. How many days are required for one crew to collect all the waste generated by the community in one week?

(A) 1 day
(B) 2 day
(C) 3 day
(D) 4 day

SITUATION FOR PROBLEM 246

A city of 23 000 people generates municipal solid waste at 1.2 kg/d·person with an as-discarded density of 134 kg/m^3. The city's collection trucks have an 18 m^3 compacted capacity for waste with a compaction density of 540 kg/m^3. Truck crews work 8 h/d, 5 d/wk, with collections occurring at each stop twice weekly. A truck crew can fill a truck in 133 min. The transfer station is centrally located and can be reached in approximately 31 min from any collection route in the city. Unloading at the transfer station requires 12 min.

246. How many trucks are required to meet the twice weekly (4 d/3 d cycle) collection schedule?

(A) one truck
(B) two trucks
(C) three trucks
(D) four trucks

SITUATION FOR PROBLEMS 247–248

The following problems address issues associated with operation of solid waste landfills.

247. Which of the following is typically not an important factor in landfill operation?

(A) limiting leachate generation
(B) controlling decomposition gas migration
(C) maintaining screening vegetation
(D) minimizing the exposed area at the active working face

248. The following measures are suggested for controlling landfill odors.

I. limiting disposal to days with little or no wind
II. applying daily cover and minimizing the exposed area at the active working face
III. applying odor-masking agents
IV. spreading the waste in shallow layers to maximize exposure to direct sunlight

Of these, which of these are most practical and effective?

(A) I and II
(B) II and III
(C) III and IV
(D) I and IV

SITUATION FOR PROBLEMS 249–250

A municipal solid waste collected from a population of 117,000 people includes 24% recycled materials that are sorted from the waste at a processing/transfer station. The population generates waste at about 3.9 lbm/day, and the waste that is not recycled is landfilled. The landfilled waste is compacted to 1100 lbm/yd^3 in a landfill with a total design capacity of 3,170,000 yd^3.

249. For a soil-cover-to-waste ratio of 1:4.5, what is the design life of the landfill?

(A) 23 yr
(B) 28 yr
(C) 35 yr
(D) 71 yr

250. What is the maximum soil-cover-to-waste ratio permissible if the desired landfill life is 25 yr?

(A) 1:1.1
(B) 1:2.5
(C) 1:5.3
(D) 1:9.8

Instructions

Name: _____
    Last   First   Middle Initial

Do not enter solutions in the test booklet. Complete solutions must be entered on the answer sheet provided by your proctor.

This is an open-book examination. You may use textbooks, handbooks, and other bound references, along with a battery-operated, silent, nonprinting calculator. Unbound reference materials and notes, scratchpaper, and writing tablets are not permitted. You may not consult with or otherwise share any materials or information with others taking the exam.

You must work all 50 multiple-choice problems in the four-hour period allocated for the afternoon session. Each of the 50 problems is worth one point. No partial credit will be awarded. Your score will be based entirely on the responses marked on the answer sheet. You may use blank spaces in the exam booklet for scratch work. However, no credit will be awarded for work shown in margins or on other pages of the exam booklet. Mark only one answer to each problem.

Principles and Practice of Engineering Examination

AFTERNOON SESSION
Sample Examination 3

251.	A	B	C	D	276.	A	B	C	D
252.	A	B	C	D	277.	A	B	C	D
253.	A	B	C	D	278.	A	B	C	D
254.	A	B	C	D	279.	A	B	C	D
255.	A	B	C	D	280.	A	B	C	D
256.	A	B	C	D	281.	A	B	C	D
257.	A	B	C	D	282.	A	B	C	D
258.	A	B	C	D	283.	A	B	C	D
259.	A	B	C	D	284.	A	B	C	D
260.	A	B	C	D	285.	A	B	C	D
261.	A	B	C	D	286.	A	B	C	D
262.	A	B	C	D	287.	A	B	C	D
263.	A	B	C	D	288.	A	B	C	D
264.	A	B	C	D	289.	A	B	C	D
265.	A	B	C	D	290.	A	B	C	D
266.	A	B	C	D	291.	A	B	C	D
267.	A	B	C	D	292.	A	B	C	D
268.	A	B	C	D	293.	A	B	C	D
269.	A	B	C	D	294.	A	B	C	D
270.	A	B	C	D	295.	A	B	C	D
271.	A	B	C	D	296.	A	B	C	D
272.	A	B	C	D	297.	A	B	C	D
273.	A	B	C	D	298.	A	B	C	D
274.	A	B	C	D	299.	A	B	C	D
275.	A	B	C	D	300.	A	B	C	D

Exam 3—Afternoon Session

SITUATION FOR PROBLEM 251

A drainage basin has the following characteristics.

total area that will discharge runoff to the detention basin	100 ha
total area of the developed site	
woodlands	40 ha
rooftops, roads, driveways, etc.	20 ha
landscaping	40 ha
runoff coefficients	
pre-developed site (woodlands)	0.25
rooftops, roads, driveways, etc.	0.95
landscaping (average slope)	0.15
25 yr rainfall intensity	9.1 cm/h
2 yr rainfall intensity	5.3 cm/h

251. Which of the following statements is true regarding the rational method for estimating runoff?

(A) It is applicable to large drainage basins.
(B) It applies to peak discharge only.
(C) It applies to drainage basins with a wide variation in site characteristics.
(D) It is more accurate for sites with more pervious surfaces.

SITUATION FOR PROBLEMS 252–255

An electroplating facility produces 250,000 gal/day of wastewater containing the following metals.

cadmium	57 mg/L as Cd^{+2}
zinc	27 mg/L as Zn^{+2}
chromium	272 mg/L as CrO_4^-

Ion exchange is used to recover the chromium and remove the cadmium and zinc from the wastewater prior to discharge. The characteristics of the ion exchangers are summarized in the following table.

characteristic	cation exchanger	anion exchanger
regenerant	H_2SO_4	NaOH
regenerant dosage, lbm/ft^3	1.2	0.45
regenerant concentration, %	1.5	2.5
regenerant density, lbm/ft^3	90	70
hydraulic loading rate, ft^3/ft^3-min	0.48	0.48
regeneration loading rate, ft^3/ft^3-min	0.18	0.18
resin capacity, equiv/L	1.6	3.4
resin bed volume, ft^3	45	45
resin bed depth, in	48	48

252. What is the regeneration period for the cation exchanger if it is to be regenerated once every 24 hr?

(A) 3.0 min
(B) 5.0 min
(C) 6.3 min
(D) 10 min

253. What is the chromic acid concentration in the recovered solution?

(A) 1.7 N
(B) 3.7 N
(C) 10 N
(D) 13 N

254. What measures could be implemented to increase HLR?

(A) add exchanger vessels in parallel
(B) add exchanger vessels in series
(C) recirculate flow
(D) none of the above

255. If the maximum feed CrO_4^- concentration to avoid resin deterioration is 0.006 N, will the feed solution require dilution?

(A) Yes, the feed concentration exceeds 0.006 N.
(B) No, the feed concentration exceeds 0.006 N.
(C) No, the feed concentration does not exceed 0.006 N.
(D) It cannot be determined from the information given.

SITUATION FOR PROBLEMS 256–261

The following questions apply to air pollution and the Clean Air Act (CAA) as amended by Congress in 1990.

256. Criteria pollutants are defined under which provision of the CAA?

(A) New Source Performance Standards (NSPS)
(B) National Ambient Air Quality Standards (NAAQS)
(C) Non-Attainment (NA) Area Regulations
(D) National Emission Standards for Hazardous Air Pollutants (NESHAP)

257. Which of the following apply to the Prevention of Significant Deterioration (PSD) program?

(A) Prioritize regional air quality as Class I, II, or III based on the sensitivity of the ecosystem.
(B) Apply best-available control technology (BACT) to any new source.
(C) Define and fix the increment of air quality degradation allowed in a region.
(D) All of the above apply.

258. What condition defines a non-attainment area?

(A) inability of a region to meet the National Emission Standards for Hazardous Air Pollutants (NESHAP)
(B) inability of a region to meet the National Ambient Air Quality Standards (NAAQS)
(C) inability of a region to meet New Source Performance Standards (NSPS)
(D) none of the above

259. What is the bubble policy for using emission reduction credits (ERC) under the emission trading program?

(A) allowing industrial expansion in non-attainment areas by letting industry compensate for new emissions by offsetting them with ERC acquired from other industries existing in the region
(B) treating all activities in a single plant or among proximate industries as a group, and allowing emissions at varied rates from the group's sources as long as the total emissions do not exceed those allowed for each separate source
(C) saving ERC for future use
(D) allowing industry to expand without acquiring a new permit as long as the net increase in emissions is not significant

260. Which of the following is not included in programs to protect the stratospheric ozone layer under Title VI of the CAA?

(A) a phase-out of production and consumption of listed substances
(B) labels on all containers used to store or transport listed substances
(C) a tax imposed on manufacturing uses of listed substances
(D) bans on the sale or distribution of nonessential products containing listed substances

261. Which of the following phrases best describes hazardous air pollutants?

(A) compounds such as particulate matter that decrease visibility
(B) compounds such as hydrocarbons and NO_x that are photochemically oxidized
(C) compounds such as asbestos and benzene that pose a risk to human health
(D) compounds such as ozone that contribute to respiratory problems

SITUATION FOR PROBLEMS 262–264

A baghouse is used by a furniture manufacturer to remove particulate from a waste air stream. The baghouse and particulate characteristics are

gas flow rate	20 m³/s
filtration velocity	0.05 m/s
air-cloth ratio	0.05 m³/m²·s
fabric resistance	0.005 atm/(cm/s)
cake resistance	0.04 atm/(g/cm²)(cm/s)
bag diameter	15 cm
bag height	2.5 m
air viscosity	2.5×10^{-4} g/cm·s
	2.5×10^{-5} kg/m·s
particulate concentration	20 g/m³
particle diameter	4.0 μm

262. What is the total fabric area required for the bag house?

(A) 40 m²
(B) 200 m²
(C) 400 m²
(D) 2000 m²

263. How many bags are required per 100 m² of total fabric area?

(A) 85
(B) 170
(C) 530
(D) 850

264. What is the pressure drop through the baghouse after 1 h of operation?

(A) 0.010 atm
(B) 0.025 atm
(C) 0.097 atm
(D) 0.15 atm

SITUATION FOR PROBLEMS 265–267

A cyclone separator is used to remove particulate from a contaminated air stream generated by a furniture manufacturer. The cyclone separator characteristics are

cyclone cylinder diameter	4.5 ft
cyclone outlet diameter	$0.75D$
cyclone cylinder length	$1.5D$
cyclone cone length	$2.5D$

The contaminated air characteristics are

flow rate	480 ft³/sec
temperature	82°F
particulate density	57 lbm/ft³

265. What is the centrifugal acceleration in the cyclone?

(A) 980 ft/sec²
(B) 1500 ft/sec²
(C) 2400 ft/sec²
(D) 4600 ft/sec²

266. What particle diameter will experience 100% removal at a centrifugal acceleration of 2000 ft/sec²?

(A) 12 μm
(B) 18 μm
(C) 20 μm
(D) 55 μm

267. What is the terminal settling velocity for the particle that experiences 100% removal?

(A) 2.2 ft/sec
(B) 3.4 ft/sec
(C) 3.1 ft/sec
(D) 14 ft/sec

SITUATION FOR PROBLEMS 268–270

An electrostatic precipitator is designed with the following characteristics.

applied voltage	65 kV
average electric field	470 000 V/m
plate width	6 m
plate length	8 m
number of plates	10
maximum specific collection area (SCA)	180 m²·s/m³
inlet particulate concentration	10 g/m³

268. What is the required plate spacing?

(A) 0.07 m
(B) 0.14 m
(C) 0.28 m
(D) 0.55 m

269. What is the gas flow rate through the precipitator for the maximum specific collector area?

(A) 0.27 m³/s
(B) 0.38 m³/s
(C) 2.7 m³/s
(D) 3.8 m³/s

270. Which of the following is generally not characteristic of electrostatic precipitators?

(A) economical
(B) efficient for moist flows and mists
(C) reliable and predictable
(D) removal efficiencies near 99%

SITUATION FOR PROBLEMS 271–272

The following illustrations present atmospheric stability conditions.

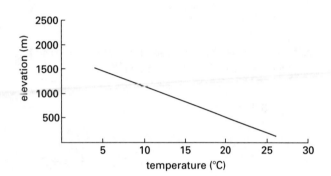

Illustration I, lapse rate = $-0.0159°C/m$

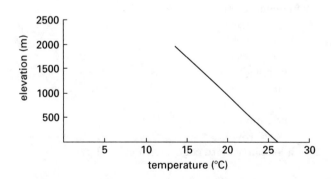

Illustration II, lapse rate = $-0.0064°C/m$

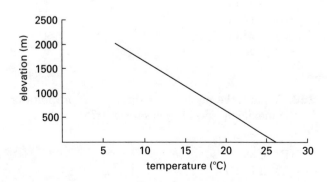

Illustration III, lapse rate = $-0.0098°C/m$

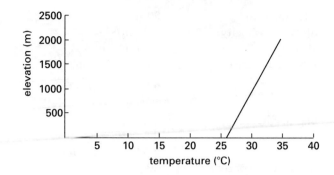

Illustration IV, lapse rate = $0.0043°C/m$

271. Which illustration depicts superadiabatic conditions?

(A) Illustration I
(B) Illustration II
(C) Illustration III
(D) Illustration IV

272. Which illustration depicts neutral conditions?

(A) Illustration I
(B) Illustration II
(C) Illustration III
(D) Illustration IV

SITUATION FOR PROBLEM 273

Emissions occur from a 359 ft high stack located in a valley with an average elevation of 3832 ft. On a particular day, an inversion layer exists at an elevation of 4114 ft. Above the inversion, subadiabatic conditions prevail.

273. What is the likely configuration of the plume?

(A) trapped on top of the inversion with vertical dispersion above
(B) trapped below the inversion with vertical dispersion beneath
(C) trapped on top of the inversion with little vertical dispersion above
(D) vertical dispersion above and below the inversion

SITUATION FOR PROBLEM 274

A plume at 31°C is emitted from a 93 m high stack where the ground-level air temperature is 19°C. The ambient lapse rate is $-0.0047°C/m$.

274. How high will the plume rise?

(A) 2170 m
(B) 2250 m
(C) 2430 m
(D) 2520 m

SITUATION FOR PROBLEMS 275–276

A pollutant is emitted from a source at 26 kg/s on a day when the wind speed is 3.2 m/s and the atmospheric stability is category C.

275. What is the ground-level concentration at 3800 m along the plume centerline for a ground-level release?

(A) 22 mg/m^3
(B) 29 mg/m^3
(C) 34 mg/m^3
(D) 38 mg/m^3

276. What is the ground-level concentration at 3800 m along the plume centerline for a release from a 110 m stack?

(A) 22 mg/m^3
(B) 29 mg/m^3
(C) 33 mg/m^3
(D) 38 mg/m^3

SITUATION FOR PROBLEMS 277–279

The following questions pertain to risk assessment associated with exposure of human populations to toxic chemicals.

277. Through what mechanism can exposure to a toxic chemical occur?

(A) through inhalation
(B) through ingestion with food or drink
(C) through contact with the skin or other exterior body surfaces
(D) through all of the above

278. In order for risk to occur, what must be present?

(A) an exposure route
(B) a toxic chemical
(C) a sensitized receptor population
(D) an exposure route and a toxic chemical

279. What four steps have traditionally defined the risk assessment process?

(A) hazard identification, dose-response assessment, exposure assessment, and risk characterization
(B) hazard identification, dose-response assessment, population assessment, and toxicity evaluation
(C) dose-response assessment, exposure assessment, risk characterization, and population assessment
(D) exposure assessment, risk characterization, population assessment, and toxicity evaluation

SITUATION FOR PROBLEMS 280–281

Children swimming in a public pool are exposed to chloroform through ingestion and dermal absorption. The following factors affect intake.

chloroform concentration	0.8 mg/L
contact rate	0.050 L/h
exposure time	2.0 h/event
exposure frequency	140 events/yr
exposure duration from age 5 yr to 12 yr	8 yr
contacted skin surface area	0.94 m^2
dermal permeability	8.4×10^{-4} cm/h

280. What is the total intake of an exposed child?

(A) 2.13×10^{-5} mg/kg·d
(B) 1.56×10^{-4} mg/kg·d
(C) 1.37×10^{-3} mg/kg·d
(D) 1.35×10^{-3} mg/kg·d

281. What is the intake of a nonswimming child from age 5 yr to 9 yr through drinking water containing trihalomethanes at the maximum contaminant level (MCL)?

(A) 1.8×10^{-4} mg/kg·d
(B) 3.5×10^{-4} mg/kg·d
(C) 1.5×10^{-3} mg/kg·d
(D) 3.1×10^{-3} mg/kg·d

SITUATION FOR PROBLEMS 282–284

The following problems address information included on a material safety data sheet (MSDS).

282. What physical data is typically not included on a MSDS?

(A) the material appearance and odor
(B) the maximum contaminant level (MCL), recommended MCL (RMCL), and MCL goal (MCLG)
(C) the molecular weight, boiling point, melting point, viscosity, and water solubility for the material
(D) the chemical formula for the material

283. What health hazard information is included on a MSDS?

(A) chemical exposure pathways into the body
(B) acute and chronic health effects
(C) medical and first aid treatments for accidental exposure
(D) all of the above

284. What information for chemical handling would not be included on a MSDS?

(A) spill response measures
(B) disposal of wastes generated from spill response
(C) decontamination of wastes generated from spill response
(D) proper storage procedures

SITUATION FOR PROBLEMS 285–288

The following problems pertain to indoor radon gas.

285. What is radon gas?

(A) a naturally occurring radioactive gas that can cause lung cancer
(B) a volatile organic compound that can cause adverse health effects
(C) a synthetic radioactive compound emitted to the atmosphere from nuclear reactor cooling water
(D) a radioactive gas characterized by pungent taste and odor

286. Which agency has regulatory authority for radon gas exposure in homes, schools, and other buildings?

(A) the Nuclear Regulatory Commission (NRC) under the Nuclear Waste Policy Act of 1982
(B) the Office of Housing and Urban Development (HUD) in association with the U.S. Department of Education under the Indoor Radon Abatement Act of 1988
(C) the U.S. Environmental Protection Agency (EPA) under the Indoor Radon Abatement Act of 1988
(D) none of the above

287. What is the standard unit of measure for radon gas?

(A) pCi/L, where pCi is the negative log of the isotope concentration per liter of air
(B) pCi/L, where pCi represents the picocuries present per liter of air
(C) ppm, representing the volume in cubic centimeters of radon gas present per standard cubic meter of air
(D) μg/L, representing the mass in micrograms of radon gas present in 1 L of water

288. What factors increase the risk of developing health problems from radon exposure?

(A) smoking, currently or in the past
(B) the level of radon existing in the home
(C) the amount of time exposed to radon gas
(D) all of the above

SITUATION FOR PROBLEMS 289–292

An underground storage tank has leaked fuel into the soil overlying an unconfined aquifer over a period of several months. The fuel exists as a light nonaqueous phase liquid (LNAPL) with the following characteristics.

density	0.811 g/cm^3
kinematic viscosity	8.32 mm^2/s

The soil-groundwater system has the following characteristics.

ambient temperature	8°C
hydraulic conductivity	19 m/d
hydraulic gradient	0.0017 m/m
effective porosity	0.24

289. What is the average velocity of the groundwater?

(A) 0.033 m/d
(B) 0.14 m/d
(C) 0.44 m/d
(D) 80 m/d

290. What is the intrinsic permeability of the soil?

(A) 2.3×10^{-13} m^2
(B) 3.1×10^{-11} m^2
(C) 2.0×10^{-8} m^2
(D) 2.7×10^{-6} m^2

291. If the intrinsic permeability of the soil is 1.0×10^{-10} m^2, what is the hydraulic conductivity with respect to the fuel?

(A) 0.020 m/d
(B) 0.27 m/d
(C) 3.2 m/d
(D) 10 m/d

292. What is the average velocity of the light nonaqueous phase fuel if the hydraulic conductivity with respect to the LNAPL is 10 m/d?

(A) 0.0054 m/d
(B) 0.023 m/d
(C) 0.071 m/d
(D) 13 m/d

SITUATION FOR PROBLEMS 293–294

The data tabulated below describe the biotransformation of 1,2,4,5-tetrachlorobenzene (1,2,4,5-TCB) in a 0.05M acetic acid wastewater.

time (h)	concentration (µg/L)
0	5.1
2	4.7
4	4.3
8	3.9
12	3.0
16	2.5
22	1.9
28	1.5

293. What is the first order reaction rate coefficient for the 1,2,4,5-TCB biodegradation?

(A) 0.024 h^{-1}
(B) 0.041 h^{-1}
(C) 0.12 µg/L·h
(D) 3.0 h·L/µg

294. Is it reasonable to expect 1,2,4,5-TCB to biodegrade to 5.1 µg/L if it exists in the wastewater as a pure solution: that is, no other compounds contribute to chemical oxygen demand (COD)?

(A) Yes, without COD, 1,2,4,5-TCB is the only available substrate.
(B) Yes, at low concentration, 1,2,4,5-TCB is easy to biodegrade.
(C) No, without COD, there are no nutrients and there is no dissolved oxygen (DO).
(D) No, even if biodegradable, 5.1 µg/L could not sustain biological activity.

SITUATION FOR PROBLEMS 295–296

Odors associated with treatment of municipal and industrial wastewaters impact siting decisions and public acceptance for treatment facilities. Odors emanating from wastewater treatment plants and from vented sanitary sewers are a major source of complaints from people who live, work, or play near areas where such facilities are located.

295. What condition is likely present when malodors from municipal wastewater treatment plants are most severe?

(A) Influent organic loading to the plant is dilute.
(B) Anaerobic conditions prevail in the bioreactor.
(C) Bulking occurs in the secondary clarifier.
(D) Primary clarification efficiency is low.

296. What compound is commonly associated with odors from wastewater treatment plants?

(A) H$_2$S gas produced from biological reduction of sulfates
(B) CO$_2$ gas produced from oxidation of organic matter
(C) CH$_4$ gas produced during anaerobic decomposition of organic matter
(D) N$_2$ produced during denitrification

SITUATION FOR PROBLEMS 297–298

The illustration presents a breakpoint chlorination curve for a water sample.

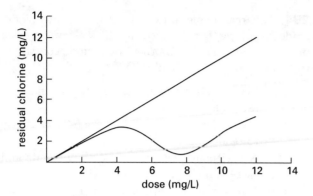

297. What is the minimum chlorine dose required to obtain a free chlorine residual?

(A) 1.5 mg/L
(B) 4.3 mg/L
(C) 7.7 mg/L
(D) 9.7 mg/L

298. What chlorine dose is required to obtain a free chlorine residual of 2.0 mg/L?

(A) 2.0 mg/L
(B) 2.2 mg/L
(C) 6.0 mg/L
(D) 9.7 mg/L

SITUATION FOR PROBLEMS 299–300

An aboveground storage tank containing a toxic volatile chemical is ruptured and 10 kg of vapor is emitted to the surrounding air. The release occurs at night with a clear sky and 2.1 m/s wind speed. The nearest residence is located 1.4 km directly downwind. The terrain is flat, with no significant obstacles present between the release area and the residence.

299. What is the maximum concentration at the residence?

(A) 11 mg/m^3
(B) 37 mg/m^3
(C) 56 mg/m^3
(D) 70 mg/m^3

300. Approximately how much time is available for the residents to evacuate their homes to avoid exposure to the vapor?

(A) < 2 min
(B) < 11 min
(C) < 25 min
(D) < 49 min

Exam 3—Solutions

#	Ans	#	Ans	#	Ans	#	Ans
201.	C	226.	D	251.	B	276.	B
202.	B	227.	C	252.	B	277.	D
203.	B	228.	B	253.	D	278.	D
204.	D	229.	B	254.	C	279.	A
205.	C	230.	C	255.	A	280.	A
206.	B	231.	C	256.	B	281.	A
207.	C	232.	A	257.	D	282.	B
208.	A	233.	B	258.	B	283.	D
209.	D	234.	C	259.	B	284.	C
210.	C	235.	D	260.	B	285.	A
211.	C	236.	B	261.	C	286.	C
212.	C	237.	C	262.	C	287.	B
213.	D	238.	C	263.	A	288.	D
214.	A	239.	A	264.	C	289.	A
215.	C	240.	A	265.	C	290.	B
216.	C	241.	D	266.	C	291.	D
217.	B	242.	B	267.	C	292.	C
218.	D	243.	A	268.	C	293.	B
219.	B	244.	C	269.	C	294.	D
220.	B	245.	B	270.	B	295.	B
221.	C	246.	B	271.	A	296.	A
222.	B	247.	C	272.	C	297.	C
223.	B	248.	B	273.	A	298.	D
224.	B	249.	A	274.	D	299.	B
225.	C	250.	D	275.	C	300.	B

201. The pre-development discharge of a 2 yr storm is

$$Q_2 = CiA$$
$$= (0.25)\left(5.3 \frac{\text{cm}}{\text{h}}\right)\left(\frac{1 \text{ m}}{100 \text{ cm}}\right)\left(\frac{1 \text{ h}}{3600 \text{ s}}\right)$$
$$\times (100 \text{ ha})\left(10\,000 \frac{\text{m}^2}{\text{ha}}\right)$$
$$= 3.7 \text{ m}^3/\text{s}$$

The answer is (C).

202.

type of surface	runoff coefficient, C	area (ha)
woodlands	0.25	40
rooftops/pavement	0.95	20
landscaping	0.15	40

$$C_{\text{ave}} = \frac{\Sigma C_i A_i}{\Sigma A_i}$$
$$= \frac{(0.25)(40 \text{ ha}) + (0.95)(20 \text{ ha}) + (0.15)(40 \text{ ha})}{100 \text{ ha}}$$
$$= 0.35$$

The post-development peak discharge of the 25 yr storm is

$$Q_{25} = (0.35)\left(9.1 \frac{\text{cm}}{\text{h}}\right)\left(\frac{1 \text{ m}}{100 \text{ cm}}\right)\left(\frac{1 \text{ h}}{3600 \text{ s}}\right)$$
$$\times (100)\left(10\,000 \frac{\text{m}^2}{\text{ha}}\right)$$
$$= 8.8 \text{ m}^3/\text{s}$$

The answer is (B).

203. The ratio of post-development to pre-development flow is 1.5:1 or 150%. For a relatively small drainage basin of 100 ha, increasing runoff by 150%, or by 0.5 m³/s, will likely considerably increase flow and justify storm water control and containment devices.

The answer is (B).

204. With no solids recycling, assume the hydraulic residence time, θ, is equal to the mean cell residence time, θ_c.

$$\frac{X\theta}{S_o - S} = \frac{K_s}{kS} + \frac{1}{k}$$

k is the maximum substrate use rate.

$\dfrac{X\theta}{S_o - S}$ (d)	$\dfrac{1}{S}$ (L/mg)
1.74	0.125
0.96	0.067
0.85	0.048
0.70	0.031
0.54	0.023

From the illustration, the intercept is

$$\frac{1}{k} = 0.29 \text{ d}$$
$$k = \frac{1}{0.29 \text{ d}}$$
$$= 3.45 \text{ d}^{-1}$$

From the illustration and the given equation, the slope is found as follows.

$$\frac{X\theta}{S_o S} = \left(\frac{K_s}{k}\right)\left(\frac{1}{S}\right) + \frac{1}{k}$$
$$\frac{K_s}{k} = \frac{1.74 \text{ d} - 0.29 \text{ d}}{0.125 \frac{\text{L}}{\text{mg}}}$$
$$= 11.6 \text{ d·mg/L}$$

The half-velocity constant is

$$K_s = \left(\frac{K_s}{k}\right)k = \left(11.6 \frac{\text{d·mg}}{\text{L}}\right)\left(\frac{3.45}{\text{d}}\right)$$
$$= 40 \text{ mg/L}$$

The answer is (D).

205. With no solids recycling, assume the hydraulic residence time, θ, is equal to the mean cell residence time, θ_c.

$$\frac{1}{\theta_c} = \frac{Y(S_o - S)}{X\theta} - k_d$$

k_d is the endogenous decay rate.

$\dfrac{1}{\theta_c}$	$\dfrac{S_o - S}{X\theta_c}$
(d^{-1})	(d^{-1})
0.29	0.57
0.53	1.04
0.63	1.18
0.77	1.44
1.0	1.86

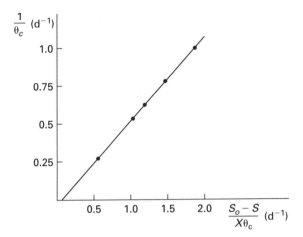

From the illustration, the slope is

$$\dfrac{\dfrac{1.0}{d} - \dfrac{0.29}{d}}{\dfrac{1.86}{d} - \dfrac{0.57}{d}} = Y$$

The maximum yield coefficient is

$$Y = 0.55$$

The answer is (C).

206.

$$\left(400 \ \dfrac{m^3}{d}\right)\left(2000 \ \dfrac{mg}{L}\right)\left(10^{-6} \ \dfrac{kg}{mg}\right)\left(10^3 \ \dfrac{L}{m^3}\right)$$
$$= 800 \ \text{kg BOD/d}$$

The pond 1 volume is

$$\dfrac{800 \ \dfrac{\text{kg BOD}}{d}}{0.35 \ \dfrac{\text{kg BOD}}{m^3 \cdot d}} = 2286 \ m^3$$

The minimum depth for pond 1 is

$$\dfrac{2286 \ m^3}{1000 \ m^2} = 2.286 \ m \quad (2.3 \ m)$$

The answer is (B).

207. Pretreatment, mixers, and aeration would likely decrease the BOD removal efficiency of pond 1 since these measures would compromise the anaerobic environment desired for the pond to operate efficiently. Only equalizing the flow to pond 1 would improve BOD removal by reducing upsets from slug loading.

The answer is (C).

208. In pond 2,

$$\left(2000 \ \dfrac{mg}{L}\right)(0.50)(0.80) = 800 \ \text{mg/L BOD removal}$$

$$\left(800 \ \dfrac{mg}{L}\right)\left(400 \ \dfrac{m^3}{d}\right)\left(10^{-6} \ \dfrac{kg}{mg}\right)$$
$$\times \left(10^3 \ \dfrac{L}{m^3}\right)\left(\dfrac{1 \ d}{24 \ h}\right)$$
$$= 13.3 \ \text{kg BOD/h}$$

$$13.3 \ \text{kg BOD/h} = 13.3 \ \text{kg DO/h}$$
$$(13 \ \text{kg DO/h})$$

$$\begin{bmatrix} \text{assuming a 1:1 ratio of dissolved} \\ \text{oxygen required for BOD removal} \end{bmatrix}$$

The answer is (A).

209. Grease facilitates formation of a scum layer that acts to insulate the pond contents from air temperature fluctuations, and that provides a barrier to aeration from surface diffusion.

The answer is (D).

The following illustration applies to Probs. 10 and 11.

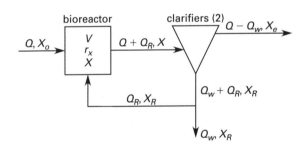

210. Perform a solids mass balance around the bioreactor. Write the statement in words.

influent VSS + return VSS
$\quad$ + cell growth = effluent VSS

Write the equation with variables.

Q $\quad$ influent flow rate $\quad$ 11 MGD
Q_R $\quad$ return solids
$\quad\quad$ flow rate $\quad$ (unknown)

r_x cell growth rate   0.028 lbm/ft³-day

V bioreactor volume $(5800 \text{ m}^3)(264 \text{ gal/m}^3)$
$\times (1 \text{ MG}/10^6 \text{ gal})$
$= 1.5 \text{ MG}$

X bioreactor VSS   2100 mg/L
X_o influent VSS   160 mg/L
X_R return VSS   14 000 mg/L

$$QX_o + Q_R X_R + V r_x = (Q + Q_R)X$$

$$\left(11 \times 10^6 \frac{\text{gal}}{\text{day}}\right)\left(160 \frac{\text{mg}}{\text{L}}\right) + Q_R \left(14\,000 \frac{\text{mg}}{\text{L}}\right)$$
$$+ (1.5 \times 10^6 \text{ gal})\left(0.028 \frac{\text{lbm}}{\text{ft}^3\text{-day}}\right)$$
$$\times \left(\frac{10^6 \text{ mg}}{2.204 \text{ lbm}}\right)\left(\frac{35.29 \text{ ft}^3}{10^3 \text{ L}}\right)$$
$$= \left(11 \times 10^6 \frac{\text{gal}}{\text{day}} + Q_R\right)$$
$$\times \left(2100 \frac{\text{mg}}{\text{L}}\right)$$

$$Q_R = 1.7 \times 10^6 \frac{\text{gal}}{\text{day}} = 1.7 \text{ MGD}$$

The answer is (C).

211. Perform a solids mass balance around the clarifier. Write the statement in words.

influent VSS = effluent VSS + return VSS
                  + wasted VSS

Write the equation with variables.

Q influent flow rate   11 MGD
Q_R return solids flow rate   1.0 MGD
Q_w wasted solids flow rate   (unknown)
X bioreactor VSS   2100 mg/L
X_e effluent VSS   30 mg/L
X_R return VSS   14 000 mg/L

$$(Q + Q_R)X = (Q + Q_w)X_e + (Q_w + Q_R)X_R$$

$$\left(11 \times 10^6 \frac{\text{gal}}{\text{day}} + 1 \times 10^6 \frac{\text{gal}}{\text{day}}\right)\left(2100 \frac{\text{mg}}{\text{L}}\right)$$
$$= \left(11 \times 10^6 \frac{\text{gal}}{\text{day}} + Q_w\right)\left(30 \frac{\text{mg}}{\text{L}}\right)$$
$$+ \left(Q_w + 1.0 \times 10^6 \frac{\text{gal}}{\text{day}}\right)\left(14\,000 \frac{\text{mg}}{\text{L}}\right)$$

$$Q_w = 7.7 \times 10^5 \frac{\text{gal}}{\text{day}} = 0.77 \text{ MGD}$$

The answer is (C).

212. Options (A), (B), and (D) would result in improved clarifier performance. Option (C), decreasing the depths of the inlet baffles, would likely contribute to hydraulic short-circuiting and result in less efficient clarifier performance.

The answer is (C).

213. Reducing the TSS concentration in the secondary clarifier effluent will likely reduce the phosphorous concentration as well, but the magnitude of the phosphorous decrease cannot be determined from the information provided.

The answer is (D).

214. The mass transfer coefficient is

$$K_L a = \frac{-\ln \frac{C}{C_o}}{t}$$

At 20 s,

$$K_L a = \frac{-\ln\left(\frac{1620 \frac{\text{mg}}{\text{L}}}{2000 \frac{\text{mg}}{\text{L}}}\right)}{20 \text{ s}} = 0.0105 \text{ s}^{-1}$$

At 60 s,

$$K_L a = \frac{-\ln\left(\frac{1023 \frac{\text{mg}}{\text{L}}}{2000 \frac{\text{mg}}{\text{L}}}\right)}{60 \text{ s}} = 0.0112 \text{ s}^{-1}$$

At 240 s,

$$K_L a = \frac{-\ln\left(\frac{134 \frac{\text{mg}}{\text{L}}}{2000 \frac{\text{mg}}{\text{L}}}\right)}{240 \text{ s}} = 0.0113 \text{ s}^{-1}$$

At 360 s,

$$K_L a = \frac{-\ln\left(\frac{45 \frac{\text{mg}}{\text{L}}}{2000 \frac{\text{mg}}{\text{L}}}\right)}{360 \text{ s}} = 0.0105 \text{ s}^{-1}$$

The overall $K_L a$ is

$$K_L a = \frac{\frac{0.0105}{\text{s}} + \frac{0.0112}{\text{s}} + \frac{0.0113}{\text{s}} + \frac{0.0105}{\text{s}}}{4}$$
$$= 0.0109 \text{ s}^{-1} \quad (0.011 \text{ s}^{-1})$$

The answer is (A).

215. Henry's constant is

$$K_H = \frac{p_g}{S_w}$$

p_g vapor pressure 0.26 atm
S_w solubility in water 9300 mg/L

$$K_H = \frac{(0.26 \text{ atm}) \left(120 \frac{\text{g}}{\text{mol}}\right) \left(\frac{\text{mg} \cdot \text{m}^3}{\text{g} \cdot \text{L}}\right)}{9300 \frac{\text{mg}}{\text{L}}}$$

$$= 0.00335 \text{ atm} \cdot \text{m}^3/\text{mol}$$

$$K_H \text{ unitless} = \frac{0.00335 \frac{\text{atm} \cdot \text{m}^3}{\text{mol}}}{\left(8.2 \times 10^{-5} \frac{\text{atm} \cdot \text{m}^3}{\text{mol} \cdot \text{K}}\right)(25°C + 273°)}$$

$$= 0.14$$

The answer is (C).

216. The air-to-water ratio is

$$\frac{V_a}{V_w} = \frac{S}{K_H} = \frac{3}{0.05} = 60$$

The answer is (C).

217. The mass-transfer coefficient, $K_L a$, can vary over a relatively wide range for a given chemical, depending on the selection of the packing material and on the temperature. A change in $K_L a$ results in a directly proportional change in the transfer unit height (HTU). The stripping factor, S, Henry's constant, K_H, and the air-to-water ratio, V_a/V_w, are related by $V_a/V_w = S/K_H$. These parameters primarily influence air flow rate and the number of transfer units (NTU). NTU does not change in direct proportion to a change in the stripping factor and will change over only a relatively narrow range.

The answer is (B).

218. Recirculating flow will not influence the mass transfer coefficient or Henry's constant, will cause a decrease in the air-to-water ratio, and will increase the hydraulic loading rate.

The answer is (D).

219.

ion	mg/L	mmol/meq	mmol/L	meq/L
SO_4^{-2}	28	2	0.29	0.58
Ca^{+2}	67	2	1.675	3.35
Mg^{+2}	21	2	0.875	1.75
HCO_3^-	353	1	5.79	5.79

$$Ca^{+2} + 2HCO_3^- + 2NaOH$$
$$\rightarrow CaCO_3 + 2Na^+ + CO_3^- + 2H_2O$$

The chemicals present are

Ca^{+2} 3.35 meq Ca^{+2}/L

HCO_3^- $\left(\frac{2 \text{ meq HCO}_3^-}{2 \text{ meq Ca}^{+2}}\right)\left(3.35 \frac{\text{meq Ca}^{+2}}{\text{L}}\right)$
$= 3.35$ meq HCO_3^-/L

The chemical required is

NaOH $\left(\frac{2 \text{ meq NaOH}}{2 \text{ meq Ca}^{+2}}\right)\left(3.35 \frac{\text{meq Ca}^{+2}}{\text{L}}\right)$
$= 3.35$ meq NaOH/L

$$Mg^{+2} + 2HCO_3^- + 4NaOH$$
$$\rightarrow Mg(OH)_2 + 4Na^+ + 2CO_3^- + 2H_2O$$

The chemicals present are

Mg^{+2} 1.75 meq Mg^{+2}/L

HCO_3^- $\left(\frac{2 \text{ meq HCO}_3^-}{2 \text{ meq Mg}^{+2}}\right)\left(1.75 \frac{\text{meq Mg}^{+2}}{\text{L}}\right)$
$= 1.75$ meq HCO_3^-/L

The chemical required is

NaOH $\left(\frac{4 \text{ meq NaOH}}{2 \text{ meq Mg}^{+2}}\right)\left(1.75 \frac{\text{meq Mg}^{+2}}{\text{L}}\right)$
$= 3.5$ meq NaOH/L

HCO_3^- remaining $= 5.79 \frac{\text{meq}}{\text{L}} - 3.35 \frac{\text{meq}}{\text{L}}$
$- 1.75 \frac{\text{meq}}{\text{L}}$
$= 0.69$ meq/L > 0 [OK]

The NaOH requirement is

$$\left(3.35 \frac{\text{meq}}{\text{L}} + 1.75 \frac{\text{meq}}{\text{L}}\right)\left(40 \frac{\text{mg}}{\text{mmol}}\right)\left(\frac{\text{mmol}}{1 \text{ meq}}\right)$$
$$= 204 \text{ mg/L}$$

$$\left(204 \frac{\text{mg}}{\text{L}}\right)\left(500 \frac{\text{gal}}{\text{min}}\right)\left(3.785 \frac{\text{L}}{\text{gal}}\right)\left(10^{-6} \frac{\text{kg}}{\text{mg}}\right)$$
$$\times \left(1440 \frac{\text{min}}{\text{d}}\right) = 556 \text{ kg/d} \quad (560 \text{ kg/d})$$

The answer is (B).

220.

ion	mg/L	mmol/meq	mmol/L	meq/L
SO_4^{-2}	28	2	0.29	0.58
Ca^{+2}	67	2	1.675	3.35
Mg^{+2}	21	2	0.875	1.75
HCO_3^-	353	1	5.79	5.79

$$Ca^{+2} + 2HCO_3^- + 2NaOH$$
$$\rightarrow CaCO_3 + 2Na^+ + CO_3^- + 2H_2O$$

The chemicals present are

Ca^{+2} 3.35 meq Ca^{+2}/L

HCO_3^- $\left(\dfrac{2 \text{ meq } HCO_3^-}{2 \text{ meq } Ca^{+2}}\right)\left(3.35 \dfrac{\text{meq } Ca^{+2}}{L}\right)$
$= 3.35$ meq HCO_3^-/L

The chemical produced is

$CaCO_3$ $\left(\dfrac{2 \text{ meq } CaCO_3}{2 \text{ meq } Ca^{+2}}\right)\left(3.35 \dfrac{\text{meq } Ca^{+2}}{L}\right)$
$= 3.35$ meq $CaCO_3$/L

$$Mg^{+2} + 2HCO_3^- + 4NaOH$$
$$\rightarrow Mg(OH)_2 + 4Na^+ + 2CO_3^- + 2H_2O$$

The chemicals present are

Mg^{+2} 1.75 meq Mg^{+2}/L

HCO_3^- $\left(\dfrac{2 \text{ meq } HCO_3^-}{2 \text{ meq } Mg^{+2}}\right)\left(1.75 \dfrac{\text{meq } Mg^{+2}}{L}\right)$
$= 1.75$ meq HCO_3^-/L

The chemical produced is

$Mg(OH)_2$ $\left(\dfrac{2 \text{ meq } Mg(OH)_2}{2 \text{ meq } Mg^{+2}}\right)\left(1.75 \dfrac{\text{meq } Mg^{+2}}{L}\right)$
$= 1.75$ meq $Mg(OH)_2$/L

HCO_3^- remaining $= 5.79 \dfrac{\text{meq}}{L} - 3.35 \dfrac{\text{meq}}{L}$
$- 1.75 \dfrac{\text{meq}}{L}$
$= 0.69$ meq/L > 0 [OK]

The sludge production is as follows.
For $CaCO_3$,

$$\left(3.35 \dfrac{\text{meq}}{L}\right)\left(100 \dfrac{\text{mg}}{\text{mmol}}\right)\left(\dfrac{1 \text{ mmol}}{2 \text{ meq}}\right) = 167.5 \text{ mg/L}$$

$$\left(167.5 \dfrac{\text{mg}}{L}\right)\left(500 \dfrac{\text{gal}}{\text{min}}\right)\left(3.785 \dfrac{L}{\text{gal}}\right)$$
$$\times \left(10^{-6} \dfrac{\text{kg}}{\text{mg}}\right)\left(1440 \dfrac{\text{min}}{d}\right) = 456 \text{ kg/d}$$

For $Mg(OH)_2$,

$$\left(1.75 \dfrac{\text{meq}}{L}\right)\left(58 \dfrac{\text{mg}}{\text{mmol}}\right)\left(\dfrac{1 \text{ mmol}}{2 \text{ meq}}\right) = 50.75 \text{ mg/L}$$

$$\left(50.75 \dfrac{\text{mg}}{L}\right)\left(500 \dfrac{\text{gal}}{\text{min}}\right)\left(3.785 \dfrac{L}{\text{gal}}\right)$$
$$\times \left(10^{-6} \dfrac{\text{kg}}{\text{mg}}\right)\left(1440 \dfrac{\text{min}}{d}\right) = 138 \text{ kg/d}$$

The total sludge is

$$456 \dfrac{\text{kg}}{d} + 138 \dfrac{\text{kg}}{d} = 594 \dfrac{\text{kg}}{d}$$

Assume sludge density is the same as water, 1000 kg/m³.
The sludge volume is

$$\dfrac{594 \dfrac{\text{kg}}{d}}{\left(1000 \dfrac{\text{kg}}{m^3}\right)(0.35)} = 1.7 \text{ m}^3/d$$

The answer is (B).

221. The volume of the settling zone is

$$V = Qt = \dfrac{\left(10000 \dfrac{m^3}{d}\right)(100 \text{ min})}{1440 \dfrac{\text{min}}{d}}$$
$$= 694.4 \text{ m}^3$$

$$A_s = \dfrac{V}{Z_o} = \dfrac{694.4 \text{ m}^3}{2.5 \text{ m}}$$
$$= 277.8 \text{ m}^2 \quad \text{[settling zone]}$$

$$A_s = lw = 3ww$$
$$= 3w^2$$

$$w = \sqrt{\dfrac{277.8 \text{ m}^2}{3}}$$
$$= 9.6 \text{ m}$$

$$l = (3)(9.6 \text{ m})$$
$$= 28.8 \text{ m} \quad \text{[settling zone]}$$

$$l_{\text{total}} = l + \text{inlet zone} + \text{outlet zone}$$
$$= 28.8 \text{ m} + 3 \text{ m} + 3 \text{ m}$$
$$= 34.8 \text{ m} \quad (35 \text{ m})$$

$$d = Z_o + \text{sludge zone depth}$$
$$= 2.5 \text{ m} + 0.5 \text{ m}$$
$$= 3.0 \text{ m}$$

The answer is (C).

222. $A_x = lZ_o = (30 \text{ m})(2.5 \text{ m}) = 75 \text{ m}^2$

The average particle-settling velocity is

$$v_h = \frac{Q}{A_x} = \frac{10\,000 \, \frac{\text{m}^3}{\text{d}}}{(75 \text{ m}^2)\left(24 \, \frac{\text{h}}{\text{d}}\right)}$$
$$= 5.6 \text{ m/h}$$

The answer is (B).

223. From the illustration at $Z_o = 2.5$ m and $t = 100$ min, $h_o = 64\%$. At $Z_o = 2.0$ m and $t = 100$ min, $h_o = 69\%$.

The efficiency increases.

The answer is (B).

224. From the illustration at $Z_o = 2.5$ m and $t = 100$ min, $h_o = 64\%$. At $Z_o = 2.5$ m and $t = 120$ min, $h_o = 77\%$.

The efficiency increases.

The answer is (B).

225. The dry mass TSS removed is

$$(0.80)\left(10\,000 \, \frac{\text{m}^3}{\text{d}}\right)\left(234 \, \frac{\text{mg}}{\text{L}}\right)\left(10^{-6} \, \frac{\text{kg}}{\text{mg}}\right)$$
$$\times \left(10^3 \, \frac{\text{L}}{\text{m}^3}\right) = 1872 \text{ kg/d} \quad [\text{dry}]$$

Assuming a solids density equal to that of water of 1000 kg/m³, the volume at 30% solids is

$$\frac{1872 \, \frac{\text{kg}}{\text{d}}}{\left(1000 \, \frac{\text{kg}}{\text{m}^3}\right)(0.30)} = 6.24 \text{ m}^3/\text{d} \quad (6.2 \text{ m}^3/\text{d})$$

The answer is (C).

226. From the illustration, the daily demand is

$$\frac{(14.9 \text{ MG} - 0 \text{ MG})\left(24 \, \frac{\text{hr}}{\text{day}}\right)}{2400 \text{ hr} - 0000 \text{ hr}} = 14.9 \text{ MGD}$$
$$(15 \text{ MGD})$$

The answer is (D).

227. The standard titrant used for alkalinity is 0.02N H_2SO_4 because 1 mL of this acid neutralizes 1 mg of alkalinity at $CaCO_3$. Therefore, because 0.03N H_2SO_4 is 1.5 times as concentrated as 0.02N H_2SO_4, 1 mL of 0.03N acid will neutralize 1.5 mg alkalinity as $CaCO_3$. This relationship is derived as follows.

The equivalent weight of $CaCO_3$ is

$$\frac{40 \, \frac{\text{g}}{\text{mol}} + 12 \, \frac{\text{g}}{\text{mol}} + (3)\left(16 \, \frac{\text{g}}{\text{mol}}\right)}{2 \, \frac{\text{equiv}}{\text{mol}}} = 50 \text{ g/equiv}$$

$$0.03\text{N } H_2SO_4 = 0.03 \, \frac{\text{equiv}}{\text{L}}$$
$$= 0.03 \text{ meq/mL}$$
$$\left(0.03 \, \frac{\text{meq}}{\text{mL}}\right)\left(50 \, \frac{\text{mg}}{\text{meq}}\right) = 1.5 \text{ mg as } CaCO_3/\text{mL}$$

The total alkalinity is

$$\frac{(\text{titrant volume})(1.5 \text{ mg alkalinity as } CaCO_3)}{(\text{sample volume})(1.0 \text{ mL titrant})}$$

$$= \frac{(28 \text{ mL})(1.5 \text{ mg alkalinity as } CaCO_3)}{(200 \text{ mL})(1.0 \text{ mL})} \times \left(1000 \, \frac{\text{mL}}{\text{L}}\right)$$

$$= 210 \text{ mg/L as } CaCO_3$$

The answer is (C).

228. The initial pH is well below the 8.3 pH system point. Therefore, essentially all alkalinity is present as bicarbonate (HCO_3^-) as illustrated.

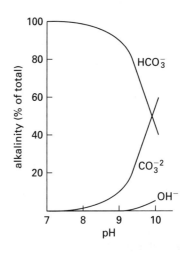

The answer is (B).

229.

ion	mg/L	mg/mmol	meq/mmol	meq/L
SO_4^{-2}	17	96	2	0.35
Na^+	48	23	1	2.1
K^+	14	39	1	0.36
Cl^-	39	35	1	1.1
Ca^{+2}	56	40	2	2.8
Mg^{+2}	12	24	2	1.0
HCO_3^-	237	61	1	3.9

The sum of the anions is

$$0.35 \, \frac{meq}{L} + 1.1 \, \frac{meq}{L} + 3.9 \, \frac{meq}{L} = 5.35 \, meq/L$$

The sum of the cations is

$$2.1 \, \frac{meq}{L} + 0.36 \, \frac{meq}{L} + 2.8 \, \frac{meq}{L} + 1.0 \, \frac{meq}{L} = 6.26 \, meq/L$$

The percent difference is

$$\left(\frac{\sum anions - \sum cations}{\sum anions + \sum cations} \right) \times 100\%$$
$$= \left(\frac{5.35 \, \frac{meq}{L} - 6.26 \, \frac{meq}{L}}{5.35 \, \frac{meq}{L} + 6.26 \, \frac{meq}{L}} \right) \times 100\%$$
$$= 7.84\%$$

Because 7.84% > 2% for the sum of the anions range between 3.0 meq/L and 10.0 meq/L, the analysis is not complete. It is deficient in anions.

The answer is (B).

230. The ions contributing to total hardness are Ca^{+2} and Mg^{+2}.

ion	mg/L	mg/mmol	meq/mmol	meq/L	mg/L as $CaCO_3$
Ca^{+2}	56	40	2	2.8	140
Mg^{+2}	12	24	2	1.0	50
					190

The total hardness is 190 mg/L as $CaCO_3$.

The answer is (C).

231. The carbonate hardness is equal to alkalinity or total hardness, whichever is least.

The total hardness is 100 mg/L as $CaCO_3$.

Assume total alkalinity is represented by HCO_3^-. The total alkalinity is

$$\frac{\left(237 \, \frac{mg}{L} \right) \left(1 \, \frac{meq}{mmol} \right) \left(50 \, \frac{mg \text{ as } CaCO_3}{meq} \right)}{61 \, \frac{mg}{mmol}}$$
$$= 194 \, mg/L \text{ as } CaCO_3$$

The carbonate hardness is equal to the total hardness, or 100 mg/L as $CaCO_3$.

The answer is (C).

232.

$$\text{noncarbonate hardness} = \text{total hardness} - \text{carbonate hardness}$$

In this case, carbonate hardness is equal to total hardness; noncarbonate hardness is equal to 0 mg/L.

The answer is (A).

233. The ratio of specific conductivity to total dissolved solids (TDS) is about 0.075, which is within the typical range when TDS is composed mostly of ionic solids.

The answer is (A).

234. Assume a 100 kg sample.

component	mass (kg)	discarded energy (kJ/kg)	discarded energy (kJ)
food	16	4650	74 400
glass	6	150	900
plastic	5	32 600	163 000
paper	44	16 750	737 000
ferrous metal	5	840	4200
non-ferrous metal	5	700	3500
yard clippings	19	6500	123 500
	100		1 106 500

$$\text{discarded energy, kJ} = (\text{mass, kg})(\text{discarded energy, kJ/kg})$$

The discarded bulk waste energy content is

$$\frac{1\,106\,500 \, kJ}{100 \, kg} = 11\,065 \, kJ/kg \quad (11\,000 \, kJ/kg)$$

The answer is (C).

235. Assume a 100 kg sample.

component	mass (kg)	moisture (%)	dry mass (kg)	discarded energy (kJ/kg)	dry energy (kJ)
food	16	70	4.80	4650	22 320
glass	6	2	5.88	150	882
plastic	5	2	4.90	32 600	159 740
paper	44	6	41.4	16 750	693 450
ferrous metal	5	3	4.85	840	4074
nonferrous metal	5	2	4.90	700	3430
yard clippings	19	60	7.60	6500	49 400
	100		74.33		933 296

$$\text{dry mass, kg} = \frac{(\text{mass, kg})(100\% - \text{moisture, \%})}{100\ \%}$$

$$\text{dry energy, kJ} = (\text{dry mass, kg}) \times (\text{discarded energy, kJ/kg})$$

The dry bulk waste energy content is

$$\frac{933\,296 \text{ kJ}}{74.33 \text{ kg}} = 12\,566 \text{ kJ/kg} \quad (13\,000 \text{ kJ/kg})$$

The answer is (D).

236.

$$\text{EBCT} = \frac{(20 \text{ m}^3)\left(1440\ \frac{\text{min}}{\text{d}}\right)}{500\ \frac{\text{m}^3}{\text{d}}}$$

$$= 57.6 \text{ min} \quad (58 \text{ min})$$

The answer is (B).

237.

$$\text{HLR} = \frac{500\ \frac{\text{m}^3}{\text{d}}}{\left(\frac{\pi}{4}\right)(3 \text{ m})^2}$$

$$= 70.7 \text{ m}^3/\text{m}^2\cdot\text{d} \quad (71 \text{ m}^3/\text{m}^2\cdot\text{d})$$

The answer is (B).

238. The initial GAC fill (virgin GAC) is

$$(2)\left(10\,000\ \frac{\text{kg}}{\text{vessel}}\right)\left(\frac{\$3.00}{\text{kg}}\right) = \$60,000$$

The replacement GAC is

$$\left(280\ \frac{\text{kg}}{\text{d}}\right)\left(365\ \frac{\text{d}}{\text{yr}}\right) = 102\,200 \text{ kg/yr}$$

The virgin GAC replacement is

$$\left(102\,200\ \frac{\text{kg}}{\text{yr}}\right)(0.15)\left(\frac{\$3.00}{\text{kg}}\right) = \$45,990$$

The reactivated replacement is

$$\left(102\,200\ \frac{\text{kg}}{\text{yr}}\right)(0.85)\left(\frac{\$2.50}{\text{kg}}\right) = \$217,175$$

The total GAC cost in the first year is

$$\$60,000 + \$45,990 + \$217,175 = \$323,165 \quad (\$320,000)$$

The answer is (C).

239. If flow was recirculated, the mass chemical loading would not change. Consequently, GAC use would not change.

The answer is (A).

240. The molar gas volume at 25°C is

$$V = \frac{R^*T}{p}$$

$$= \frac{\left(8.2 \times 10^{-5}\ \frac{\text{atm}\cdot\text{m}^3}{\text{mol}\cdot\text{K}}\right)(25°\text{C} + 273°)}{(1 \text{ atm})\left(10^{-3}\ \frac{\text{m}^3}{\text{L}}\right)}$$

$$= 24.4 \text{ L/mol}$$

$$CO_2 \text{ MW} = 12\ \frac{\text{g}}{\text{mol}} + (2)\left(16\ \frac{\text{g}}{\text{mol}}\right)$$

$$= 44 \text{ g/mol}$$

$$(0.25)\left(60,000\ \frac{\text{ft}^3}{\text{day}}\right)\left(\frac{1 \text{ mol}}{24.4 \text{ L}}\right)\left(28.2\ \frac{\text{L}}{\text{ft}^3}\right)$$

$$= 17\,336 \text{ mol/d}$$

$$m = \left(17\,336 \, \frac{\text{mol}}{\text{d}}\right)\left(44 \, \frac{\text{g}}{\text{mol}}\right)\left(\frac{1 \text{ kg}}{1000 \text{ g}}\right)$$
$$= 763 \text{ kg/d} \quad (760 \text{ kg/d})$$

The answer is (A).

241. Henry's constant, K_H, for CO_2 at 10°C is 1040 atm/mol fraction.

Assume all the CO_2 is used when carbonating.

$$C_{eq} = \frac{\rho_g}{K_H}$$

C_{eq} CO_2 concentration (unknown)
ρ_g partial pressure 1.2 atm

$$C_{eq} = \frac{1.2 \text{ atm}}{1040 \, \frac{\text{atm}}{\text{mol fraction}}}$$
$$= 0.00115 \text{ mol fraction}$$

$$0.00115 \text{ mol fraction} = \frac{\text{mol } CO_2}{\text{mol } CO_2 + \text{mol } H_2O}$$

$$\text{mol } H_2O = \frac{1000 \, \frac{\text{g}}{\text{L}}}{16 \, \frac{\text{g}}{\text{mol}} + (2)\left(1 \, \frac{\text{g}}{\text{mol}}\right)}$$
$$= 55.6 \text{ mol/L}$$

In 1 L of carbonated water,

$$\frac{\text{mol } CO_2}{\text{L}} = 0.00115 \, \frac{\text{mol}}{\text{L}} + (0.00115)\left(55.6 \, \frac{\text{mol}}{\text{L}}\right)$$
$$= 0.065 \text{ mol/L}$$

$$\left(0.065 \, \frac{\text{mol}}{\text{L}}\right)\left(12 \, \frac{\text{g}}{\text{mol}} + (2)\left(16 \, \frac{\text{g}}{\text{mol}}\right)\right) = 2.9 \text{ g/L}$$

$$\frac{(100 \text{ kg})\left(1000 \, \frac{\text{g}}{\text{kg}}\right)}{2.9 \, \frac{\text{g}}{\text{L}}} = 34\,483 \text{ L/100 kg}$$
$$(35\,000/100 \text{ kg})$$

The answer is (D).

242. $$Pb^{+2} + 2OH^- \rightarrow Pb(OH)_2$$

$\Delta G°$ Pb^{+2} = -5.83 kcal/mol
$\Delta G°$ OH^- = -37.6 kcal/mol
$\Delta G°$ $Pb(OH)_2$ = -108.1 kcal/mol

$$\Delta G°_{reaction} = \sum \Delta G°_{products} - \sum \Delta G°_{reactants}$$
$$= (1 \text{ mol})\left(-108.1 \, \frac{\text{kcal}}{\text{mol}}\right)$$
$$- (1 \text{ mol})\left(-5.83 \, \frac{\text{kcal}}{\text{mol}}\right)$$
$$- (2 \text{ mol})\left(-37.6 \, \frac{\text{kcal}}{\text{mol}}\right)$$
$$= -27.07 \text{ kcal/mol}$$

The negative sign indicates the reaction proceeds as written. Therefore, sodium hydroxide will precipitate Pb(II).

The answer is (B).

243. The solubility product for lead hydroxide at 25°C is
$$K_{sp} = 2.5 \times 10^{-16}$$
$$\ln\left(\frac{K_{sp43}}{K_{sp25}}\right) = \frac{\Delta H°_{reaction}(T_{25} - T_{43})}{RT_{25}T_{43}}$$

R is the ideal gas constant.

At 25°C,

$\Delta H°$ Pb^{+2} = -0.4 kcal/mol
$\Delta H°$ OH^- = -54.97 kcal/mol
$\Delta H°$ $Pb(OH)_2$ = -123.3 kcal/mol

$$\Delta H°_{reaction} = \Delta H°_{products} - \Delta H°_{reactants}$$
$$= (1 \text{ mol})\left(-123.3 \, \frac{\text{kcal}}{\text{mol}}\right)$$
$$- (1 \text{ mol})\left(-0.4 \, \frac{\text{kcal}}{\text{mol}}\right)$$
$$- (2 \text{ mol})\left(-54.97 \, \frac{\text{kcal}}{\text{mol}}\right)$$
$$= -12.96 \text{ kcal/mol}$$

$$\ln\left(\frac{K_{sp43}}{K_{sp25}}\right) = \frac{\left(12.96 \, \frac{\text{kcal}}{\text{mol}}\right)(298\text{K} - 316\text{K})}{\left(1.99 \times 10^{-3} \, \frac{\text{kcal}}{\text{mol·K}}\right)(298\text{K})(316\text{K})}$$
$$= -1.245$$

$$\frac{K_{sp43}}{K_{sp25}} = e^{-1.245} = 0.288$$
$$K_{sp43} = 0.288 K_{sp25}$$
$$= (0.288)(2.5 \times 10^{-16})$$
$$= 7.2 \times 10^{-17}$$

The answer is (A).

244. $Pb^{+2} + OH^- \rightarrow PbOH^+$

k_i are stepwise formation constants.

$$\frac{[PbOH^+]}{[Pb^{+2}][OH^-]} = k_1 = 10^{7.82}$$

$$PbOH^+ + OH^- \rightarrow Pb(OH)_2$$

$$\frac{[Pb(OH)_2]}{[PbOH^+][OH^-]} = k_2 = 10^{3.03}$$

$$Pb(OH)_2 + OH^- \rightarrow Pb(OH)_3^-$$

$$\frac{[Pb(OH)_3^-]}{[Pb(OH)_2][OH^-]} = k_3 = 10^{3.73}$$

The molar concentration of Pb^{+2} is

$$\frac{\left(72 \frac{mg}{L}\right)\left(10^{-3} \frac{g}{mg}\right)}{207 \frac{g}{mol}} = 0.000\,348 \text{ mol/L}$$

The molar concentration of OH^- at pH 7.6 is by definition

$$pH + pOH = 14$$
$$pOH = 14 - 7.6$$
$$= 6.4$$
$$[OH^-] = 10^{-pOH}$$
$$[OH^-] = 10^{-6.4} \frac{mol}{L}$$
$$= 3.98 \times 10^{-7} \text{ mol/L}$$

$$[PbOH^+] = (10^{7.82})\left(0.000\,348 \frac{mol}{L}\right)$$
$$\times \left(3.98 \times 10^{-7} \frac{mol}{L}\right)$$
$$= 0.009\,15 \text{ mol/L}$$

$$\left(0.009\,15 \frac{mol}{L}\right)\left(207 \frac{g}{mol}\right)\left(1000 \frac{mg}{g}\right)$$
$$= 1894 \text{ mg/L as } Pb^{+2}$$

$$[Pb(OH)_2] = (k_2)[pbOH^+][OH^-]$$
$$= (10^{3.03})\left(0.009\,15 \frac{mol}{L}\right)$$
$$\times \left(3.98 \times 10^{-7} \frac{mol}{L}\right)$$
$$= 3.9 \times 10^{-6} \text{ mol/L}$$

$$\left(3.9 \times 10^{-6} \frac{mol}{L}\right)\left(207 \frac{g}{mol}\right)\left(1000 \frac{mg}{g}\right)$$
$$= 0.808 \text{ mg/L as } Pb^{+2}$$

$$[Pb(OH)_3^-] = (k_3)[Pb(OH)_2][OH^-]$$
$$= (10^{3.73})\left(3.9 \times 10^{-6} \frac{mol}{L}\right)$$
$$\times \left(3.98 \times 10^{-7} \frac{mol}{L}\right)$$
$$= 8.3 \times 10^{-9} \text{ mol/L}$$

$$\left(8.3 \times 10^{-9} \frac{mol}{L}\right)\left(207 \frac{g}{mol}\right)\left(1000 \frac{mg}{g}\right)$$
$$= 0.0017 \text{ mg/L as } Pb^{+2}$$

At pH 7.6, the dominant Pb^{+2} specie is $PbOH^+$.

The answer is (B).

245. The total as-discarded waste volume to be collected is

$$\frac{(1400 \text{ people})\left(3.6 \frac{lbm}{day \cdot person}\right)}{240 \frac{lbm}{yd^3}} = 21 \text{ yd}^3/\text{day}$$

Assume waste collection occurs once each week. The compacted volume per week is

$$\frac{\left(21 \frac{yd^3}{day}\right)\left(7 \frac{days}{wk}\right)}{\left(\frac{3.1 \text{ yd}^3 \text{ discarded}}{1 \text{ yd}^3 \text{ compacted}}\right)} = 47.4 \text{ yd}^3/\text{wk}$$

The number of truck loads to collect all the waste in one week is

$$\frac{47.4 \frac{yd^3}{wk}}{10 \frac{yd^3}{load}} = 4.74 \text{ load/wk} \quad (5 \text{ load/wk})$$

The time required to fill one truck is

$$\frac{\left(\frac{32 \text{ sec} + 18 \text{ sec}}{\text{residence}}\right)\left(10 \frac{yd^3}{load}\right)}{\left(\frac{47.4 \text{ yd}^3}{388 \text{ residences}}\right)\left(60 \frac{sec}{min}\right)} = 68 \text{ min}$$

The available collection time is

$$\left(8 \frac{hr}{day}\right)\left(60 \frac{min}{hr}\right) = 480 \text{ min/day}$$

At the beginning of day 1,

activity	incremental time (min)	cumulative time (min)
travel to route	28	28
collect load 1	68	96
travel to landfill	47	149
unload	15	158
travel to route	47	205
collect load 2	68	273
travel to landfill	47	320
unload	15	335
travel to route	47	382
collect load 3	68	450
travel to yard	28	478

Compared to the total time available of 480 min, the cumulative time for all activities of 478 min is 2 min early.

At the beginning of day 2,

activity	incremental time (min)	cumulative time (min)
travel to landfill	34	34
unload	15	49
travel to route	47	96
collect load 4	68	164
travel to landfill	47	211
unload	15	226
travel to route	47	273
collect load 5	68	341
travel to landfill	47	388
unload	15	403
travel to yard	34	437

Compared to the total time available of 480 min, the cumulative time for all activities of 437 min is 43 min early.

The number of days required to collect all the waste is two.

The answer is (B).

246. The total weekly compacted volume requiring collection is

$$\frac{(23\,000 \text{ people})\left(1.2 \frac{\text{kg}}{\text{person} \cdot \text{d}}\right)\left(7 \frac{\text{d}}{\text{wk}}\right)}{540 \frac{\text{kg}}{\text{m}^3}} = 358 \text{ m}^3/\text{wk}$$

The volume collected during the first collection period covering 4 d is

$$\left(\frac{4 \text{ d}}{7 \text{ d}}\right)\left(358 \frac{\text{m}^3}{\text{wk}}\right) = 205 \text{ m}^3/\text{wk}$$

The volume collected during the second collection period covering 3 d is

$$\left(\frac{3 \text{ d}}{7 \text{ d}}\right)\left(358 \frac{\text{m}^3}{\text{wk}}\right) = 153 \text{ m}^3/\text{wk}$$

The time available for collection is

$$\left(8 \frac{\text{h}}{\text{d}}\right)\left(60 \frac{\text{min}}{\text{h}}\right) = 480 \text{ min}$$

activity	incremental time (min)	cumulative time (min)
travel to route	31	31
collect load 1	133	164
return to transfer station	31	195
unload	12	207
travel to route	31	238
collect load 2	133	371
return to transfer station	31	402
unload	12	414

Insufficient time remains to collect a third full load. A single truck can complete two trips in one day.

For the first collection period, the required number of trucks is

$$\frac{\left(205 \frac{\text{m}^3}{\text{wk}}\right)\left(\frac{1 \text{ wk}}{5 \text{ d}}\right)\left(\frac{1 \text{ d}}{2 \text{ loads}}\right)}{18 \frac{\text{m}^3}{\text{load}}} = 1.14 \text{ trucks}$$

If a third partial load is collected, the time available for collection is as follows.

activity	incremental time (min)	cumulative time (min)
unload	12	414
collect partial load 3	(unknown)	
return to transfer station	31	445
unload	12	457

$$480 \text{ min} - 457 \text{ min} = 23 \text{ min}$$

The volume of partial load 3 is

$$(18 \text{ m}^3)\left(\frac{23 \text{ min}}{133 \text{ min}}\right) = 3.1 \text{ m}^3$$

The volume of 0.14 truck load is

$$(0.14)(18 \text{ m}^3) = 2.5 \text{ m}^3 < 3.1 \text{ m}^3$$

Completing partial load 3 will allow one truck to collect all the waste generated during the first collection period.

For the second collection period, the required number of trucks is

$$\frac{\left(153 \; \frac{m^3}{wk}\right)\left(\frac{1 \; wk}{5 \; d}\right)\left(\frac{1 \; d}{2 \; loads}\right)}{18 \; \frac{m^3}{load}} = 0.85 \; trucks$$

The total number of trucks required is two.

The answer is (B).

247. Modern landfills are designed, constructed, and operated to limit leachate generation and contain any leachate that is generated to prevent its migration outside the landfill boundaries, to control decomposition gas migration by providing collection and interception devices, and to minimize the area exposed as an active face by applying daily cover to minimize odors, scattered debris, and disease-causing vectors. Maintaining screening vegetation may be necessary at some locations, but it is typically not included as an important factor in landfill operation.

The answer is (C).

248. Landfill odors are effectively controlled by applying daily cover and minimizing the exposed area at the active working face. In some cases, chemical masking agents may also be applied to help suppress odors. It is usually not practical to limit landfill operation to days with little or no wind unless unfavorable conditions occur only very rarely. Also, spreading waste in shallow layers is contrary to the current landfill practice of promptly compacting and covering waste and may lead to dispersion of debris by wind, increase access for disease-causing vectors, as well as exacerbate odor problems.

The answer is (B).

249. The annual waste volume to be landfilled is

$$\frac{(1 - 0.24)(117{,}000 \; \text{people})}{1100 \; \frac{lbm}{yd^3}} \times \left(3.9 \; \frac{lbm}{day\text{-}person}\right)\left(365 \; \frac{days}{yr}\right) = 115{,}071 \; yd^3/yr$$

The annual cover volume is

$$\frac{\left(115{,}071 \; \frac{yd^3}{yr}\right)}{4.5} = 25{,}571 \; yd^3/yr$$

The design life of the landfill is

$$\frac{3{,}170{,}000 \; yd^3}{115{,}071 \; \frac{yd^3}{yr} + 25{,}571 \; \frac{yd^3}{yr}} = 22.54 \; yr \quad (23 \; yr)$$

The answer is (A).

250. The annual waste volume to be landfilled is

$$\frac{(1 - 0.24)(117{,}000 \; \text{people})}{1100 \; \frac{lbm}{yd^3}} \times \left(3.9 \; \frac{lbm}{day\text{-}person}\right)\left(365 \; \frac{days}{yr}\right) = 115{,}071 \; yd^3/yr$$

The maximum soil-cover-to-waste ratio, one part soil to x parts waste, is

$$\frac{3{,}170{,}000 \; yd^3}{115{,}071 \; \frac{yd^3}{yr} + \frac{115{,}071 \; \frac{yd^3}{yr}}{x}} = 25 \; yr$$

$$x = 9.81 \quad (9.8)$$

The ratio is 1:9.8.

The answer is (D).

251. The rational method is applicable to small drainage basins with homogeneous site conditions and impervious surfaces. It applies to peak discharge only.

The answer is (B).

252. In the cation exchanger, Cd^{+2} and Zn^{+2} are exchanged for H^+.

The concentration of cations to be removed is

$$\frac{\left(57 \; \frac{mg \; Cd}{L}\right)\left(2 \; \frac{equiv}{mol}\right)}{\left(112 \; \frac{g \; Cd}{mol}\right)\left(1000 \; \frac{mg}{g}\right)} + \frac{\left(27 \; \frac{mg \; Zn}{L}\right)\left(2 \; \frac{equiv}{mol}\right)}{\left(65 \; \frac{g \; Zn}{mol}\right)\left(1000 \; \frac{mg}{g}\right)}$$

$$= 0.00185 \; equiv/L$$

The resin volume used during 1 day of operation is

$$\frac{\left(250{,}000 \; \frac{gal}{day}\right)\left(0.00185 \; \frac{equiv}{L}\right)\left(0.134 \; \frac{ft^3}{gal}\right)}{1.6 \; \frac{equiv}{L}}$$

$$= 38.7 \; ft^3/day$$

The regeneration volume is

$$\frac{\left(1.2 \ \frac{\text{lbm}}{\text{ft}^3}\right)\left(38.7 \ \frac{\text{ft}^3}{\text{day}}\right)}{\left(90 \ \frac{\text{lbm}}{\text{ft}^3}\right)\left(\frac{1.5 \ \text{lbm H}_2\text{SO}_4}{100 \ \text{lbm solution}}\right)} = 34.4 \ \text{ft}^3/\text{day}$$

The regeneration period is

$$\frac{34.4 \ \frac{\text{ft}^3}{\text{day}}}{\left(0.18 \ \frac{\text{ft}^3}{\text{ft}^3\text{-min}}\right)\left(38.7 \ \frac{\text{ft}^3}{\text{day}}\right)} = 4.94 \ \text{min} \quad (5.0 \ \text{min})$$

The answer is (B).

253. In the anion exchanger, CrO_4^{2-} is exchanged for OH^-.

The concentration of CrO_4^- to be removed is

$$\frac{\left(272 \ \frac{\text{mg CrO}_4^-}{\text{L}}\right)\left(6 \ \frac{\text{equiv}}{\text{mol}}\right)\left(10^{-3} \ \frac{\text{g}}{\text{mg}}\right)}{52 \ \frac{\text{g}}{\text{mol}} + (4)\left(16 \ \frac{\text{g}}{\text{mol}}\right)}$$

$$= 0.0141 \ \text{equiv/L}$$

The resin volume used during 1 day of operation is

$$\frac{\left(250{,}000 \ \frac{\text{gal}}{\text{day}}\right)\left(0.0141 \ \frac{\text{equiv}}{\text{L}}\right)\left(0.134 \ \frac{\text{ft}^3}{\text{gal}}\right)}{\left(3.4 \ \frac{\text{equiv}}{\text{L}}\right)}$$

$$= 139 \ \text{ft}^3/\text{day}$$

The regeneration volume is

$$\frac{\left(0.45 \ \frac{\text{lbm}}{\text{ft}^3}\right)\left(139 \ \frac{\text{ft}^3}{\text{day}}\right)}{\left(70 \ \frac{\text{lbm}}{\text{ft}^3}\right)\left(\frac{2.5 \ \text{lbm NaOH}}{100 \ \text{lbm solution}}\right)} = 36 \ \text{ft}^3/\text{day}$$

The recovered CrO_4^- concentration is

$$\frac{\left(0.0141 \ \frac{\text{equiv}}{\text{L}}\right)\left(250{,}000 \ \frac{\text{gal}}{\text{day}}\right)}{\left(36 \ \frac{\text{ft}^3}{\text{day}}\right)\left(7.46 \ \frac{\text{gal}}{\text{ft}^3}\right)} = 13.1 \ \text{equiv/L}$$

$$= 13.1 \ \text{N} \quad (13 \ \text{N})$$

The answer is (D).

254. The hydraulic loading rate can be increased by increasing the flow to the exchangers or by decreasing the exchanger bed area. Bed area is typically fixed by standard vessel sizing, but flow can be increased by recirculation.

The answer is (C).

255. The feed concentration of CrO_4^- is

$$\frac{\left(272 \ \frac{\text{mg}}{\text{L}}\right)\left(6 \ \frac{\text{equiv}}{\text{mol}}\right)\left(10^{-3} \ \frac{\text{g}}{\text{mg}}\right)}{52 \ \frac{\text{g}}{\text{mol}} + (4)\left(16 \ \frac{\text{g}}{\text{mol}}\right)} = 0.0141 \ \frac{\text{equiv}}{\text{L}}$$

$$= 0.0141 \ \text{N}$$

$$0.0141 \ \text{N} > 0.006 \ \text{N}$$

The feed solution requires dilution.

The answer is (A).

256. Criteria pollutants are defined in the CAA under the National Ambient Air Quality Standards (NAAQS).

The answer is (B).

257. The PSD program prioritizes regions as Class I, II, or III, depending on the sensitivity of the ecosystem, with Class I being the highest priority. Any new source may not contribute to the significant deterioration of the air quality and is required to apply the best-available control technology (BACT). Also, any increment of air quality degradation that may be allowed is defined and fixed.

The answer is (D).

258. A nonattainment area is defined as a region where the National Ambient Air Quality Standards (NAAQS) cannot be met.

The answer is (B).

259. The bubble policy allows industry flexibility by treating all activities in a single plant or among a group of proximate industries to emit at various rates as long as the resulting total emissions do not exceed the allowable emissions for each individual source.

The answer is (B).

260. Current international law closely regulates the production and sale of chlorofluorocarbons (CFCs)—listed substances thought to contribute to ozone layer destruction—through bans on their use and production, taxes on emissions from their use, and a permit program.

The answer is (B).

261. Hazardous air pollutants are limited to asbestos, inorganic arsenic, benzene, mercury, beryllium, radionuclides, radon-222, and vinyl chloride.

The answer is (C).

262. The total fabric area is

$$\dfrac{Q_g}{\dfrac{A}{C}} = \dfrac{20 \, \dfrac{m^3}{s}}{0.05 \, \dfrac{m^3}{m^2 \cdot s}} = 400 \text{ m}^2$$

The answer is (C).

263. The area per bag is

$$\begin{aligned}\pi(\text{bag diameter}) \\ \times (\text{bag length})\end{aligned} = \pi(15 \text{ cm})\left(\dfrac{1 \text{ m}}{100 \text{ cm}}\right)(2.5 \text{ m})$$
$$= 1.18 \text{ m}^2/\text{bag}$$

The number of bags per 100 m² of total bag area is

$$\dfrac{100 \text{ m}^2}{1.18 \, \dfrac{m^2}{\text{bag}}} = 84.75 \text{ bags} \quad (85 \text{ bags})$$

The answer is (A).

264. The pressure drop is

$$-\Delta P = K_1 v_f + K_2 C_p v_f^2 t$$
$$= \dfrac{(0.005 \text{ atm})\left(0.05 \, \dfrac{m}{s}\right)}{\left(1 \, \dfrac{cm}{s}\right)\left(\dfrac{1 \text{ m}}{100 \text{ cm}}\right)}$$
$$+ \dfrac{(0.04 \text{ atm})\left(20 \, \dfrac{g}{m^3}\right)\left(0.05 \, \dfrac{m}{s}\right)^2(1 \text{ h})}{\left(1 \, \dfrac{g}{cm^2}\right)\left(1 \, \dfrac{cm}{s}\right)\left(100 \, \dfrac{cm}{m}\right)\left(\dfrac{1 \text{ h}}{3600 \text{ s}}\right)}$$
$$= 0.097 \text{ atm}$$

The answer is (C).

265. The average cyclone diameter is

$$D_a = D_e + 0.5(D - D_e)$$
$$= (0.75)(4.5 \text{ ft}) + (0.5)(4.5 \text{ ft} - (0.75)(4.5 \text{ ft}))$$
$$= 3.94 \text{ ft}$$

The average cyclone radius is

$$r_a = \dfrac{3.94 \text{ ft}}{2} = 1.97 \text{ ft}$$

The centrifugal acceleration in the cyclone, g_c, is determined from

$$\sqrt{g_c r_a} = \dfrac{4Q_g}{(D - D_e)(2L_1 + L_2)}$$
$$= \dfrac{(4)\left(480 \, \dfrac{ft^3}{\sec}\right)}{(4.5 \text{ ft} - (0.75)(4.5 \text{ ft}))\left(\begin{array}{c}(2)(1.5)(4.5 \text{ ft})\\ + (2.5)(4.5 \text{ ft})\end{array}\right)}$$
$$= 68.96 \text{ ft/sec}$$

$$g_c = \dfrac{\left(68.96 \, \dfrac{ft}{\sec}\right)^2}{1.97 \text{ ft}}$$
$$= 2414 \text{ ft/sec}^2 \quad (2400 \text{ ft/sec}^2)$$

The answer is (C).

266. At 82°F, the absolute viscosity, μ_g, of air is 1.26×10^{-5} lbm/ft sec.

The diameter of the 100% removed particle is

$$d_p = 4\sqrt{\dfrac{Q_g \mu_g}{g_c \rho_p (2L_1 + L_2)(D + D_e)}}$$

$$= 4\sqrt{\dfrac{\left(480 \, \dfrac{ft^3}{\sec}\right)\left(1.26 \times 10^{-5} \, \dfrac{lbm}{ft\text{-}sec}\right)}{\left(2000 \, \dfrac{ft}{\sec^2}\right)\left(57 \, \dfrac{lbm}{ft^3}\right)\left(\begin{array}{c}(2)(1.5)(4.5 \text{ ft})\\ + (2.5)(4.5 \text{ ft})\end{array}\right)}}$$
$$\times \left(\begin{array}{c}4.5 \text{ ft} + (0.75)\\ \times (4.5 \text{ ft})\end{array}\right)$$
$$\times \left(\dfrac{10^6 \, \mu m}{3.28 \text{ ft}}\right)$$
$$= 20.1 \, \mu m \quad (20 \, \mu m)$$

The answer is (C).

267. The terminal settling velocity of the 100% removed particle is

$$v_t = \dfrac{Q_g}{0.5\pi(L_1 + 0.5 L_2)(D + D_e)}$$
$$= \dfrac{480 \, \dfrac{ft^3}{\sec}}{\begin{array}{c}0.5\pi((1.5)(4.5 \text{ ft}) + (0.5)(2.5)(4.5 \text{ ft}))\\ \times (4.5 \text{ ft} + (0.75)(4.5 \text{ ft}))\end{array}}$$
$$= 3.1 \text{ ft/sec}$$

The answer is (C).

268. The distance between discharge electrodes and the collectors is

$$Z_e = \dfrac{V_d}{E_f} = \dfrac{(65 \text{ kV})\left(1000 \, \dfrac{V}{kV}\right)}{470\,000 \, \dfrac{V}{m}}$$
$$= 0.138 \text{ m}$$

The distance between collectors is

$$(2)(0.138 \text{ m}) = 0.276 \text{ m} \quad (0.28 \text{ m})$$

The answer is (C).

269. The total collector plate area is

$$A = (6 \text{ m})(8 \text{ m})(10) = 480 \text{ m}^2$$

The gas flow rate based on maximum specific collector area is

$$Q_g = \frac{A}{\text{SCA}} = \frac{480 \text{ m}^2}{180 \frac{\text{m}^2 \cdot \text{s}}{\text{m}^3}}$$

$$= 2.7 \text{ m}^3/\text{s}$$

The answer is (C).

270. Electrostatic precipitators are used to remove particulate matter from gas streams. They are popular because they are economical to operate, provide high removal efficiencies (near 99%), are dependable and predictable, and do not produce a moisture plume. Electrostatic precipitators generally cannot be used with moist flows and mists, since performance is inhibited by water droplets that can insulate particles and reduce their resistivities.

The answer is (B).

271. The dry adiabatic lapse rate is $-0.0098°\text{C/m}$ and is represented by Illustration III. Superadiabatic conditions occur when the temperature increases with elevation at a slower rate than dictated by the dry adiabatic lapse rate as shown in Illustration I.

The answer is (A).

272. Conditions depicted by Illustration III are neutral, those shown by illustrations II and IV are stable, and those shown by Illustration I are unstable.

The answer is (C).

273. The stack height will emit the plume above the inversion, and subadiabatic conditions will allow the plume to disperse vertically upward. The likely plume configuration will have the plume trapped on top of the inversion with vertical dispersion above.

The answer is (A).

274. The adiabatic lapse rate is $0.0098°\text{C/m}$.

The air temperature at 93 m is

$$19°\text{C} + \left(-0.0047° \frac{\text{C}}{\text{m}}\right)(93 \text{ m}) = 18.6°\text{C}$$

$$31°\text{C} + \left(-0.0098° \frac{\text{C}}{\text{m}}\right)z = 18.6°\text{C} + \left(-0.0047° \frac{\text{C}}{\text{m}}\right)z$$

The height above the stack where plume and air temperatures are equal is

$$z = \frac{31°\text{C} - 18.6°\text{C}}{0.0098° \frac{\text{C}}{\text{m}} - 0.0047° \frac{\text{C}}{\text{m}}} = 2431 \text{ m}$$

The plume will rise to

$$2431 \text{ m} + 93 \text{ m} = 2524 \quad (2520 \text{ m})$$

The answer is (D).

275. Using published vertical dispersion coefficient plots, for a 3800 m distance the standard deviation along the vertical axis is $\sigma_z = 200$ m.

Using published horizontal dispersion coefficient plots, for a 3800 m distance the standard deviation along the horizontal axis is $\sigma_y = 380$ m.

Q_g emission rate 26 kg/s
μ wind speed 3.2 m/s

The plume centerline concentration at 3800 m downwind is

$$C_{3800,0} = \frac{Q_g}{\pi \mu \sigma_z \sigma_y}$$

$$= \frac{\left(26 \frac{\text{kg}}{\text{s}}\right)\left(10^6 \frac{\text{mg}}{\text{kg}}\right)}{\pi \left(3.2 \frac{\text{m}}{\text{s}}\right)(200 \text{ m})(380 \text{ m})}$$

$$= 34 \text{ mg/m}^3$$

The answer is (C).

276. Using published vertical dispersion coefficient plots, the standard deviation along the vertical axis for a 3800 m distance is $\sigma_z = 200$ m.

Using published horizontal dispersion coefficient plots, the standard deviation along the horizontal axis for a 3800 m distance is $\sigma_y = 380$ m.

Q_g emission rate 26 kg/s
μ wind speed 3.2 m/s

The plume centerline concentration at 3800 m downwind is

$$C_{3800,0} = \frac{Q_g e^{-0.5(H/\sigma_z)^2}}{\pi \mu \sigma_z \sigma_y}$$

$$= \frac{\left(26 \, \dfrac{\text{kg}}{\text{s}}\right)\left(10^6 \, \dfrac{\text{mg}}{\text{kg}}\right) e^{(-0.5)(110 \text{ m}/200 \text{ m})^2}}{\pi \left(3.2 \, \dfrac{\text{m}}{\text{s}}\right)(200 \text{ m})(380 \text{ m})}$$

$$= 29.3 \text{ mg/m}^3 \quad (29 \text{ mg/m}^3)$$

The answer is (B).

277. Exposure to a toxic chemical can occur through inhalation, ingestion with food or drink, and contact with the skin and other exterior body surfaces.

The answer is (D).

278. For risk to occur, an exposure route and a toxic chemical must be present.

The answer is (D).

279. The traditional four steps of the risk assessment process include hazard identification, dose-response assessment, exposure assessment, and risk characterization.

The answer is (A).

280. Chloroform is a carcinogen. The EPA recommended values for estimating intake are

average time of exposure for carcinogens,
$$\text{AT} = (70 \text{ yr})\left(365 \, \dfrac{\text{d}}{\text{yr}}\right) = 25\,550 \text{ d}$$

body weight for a child aged 5 yr to 12 yr,
$$\text{BW} = 26 \text{ kg}$$

The intake through ingestion while swimming is

$$I_I = \frac{CR_c t_E f_E D_t}{(\text{BW})(\text{AT})}$$

$$= \frac{\left(0.8 \, \dfrac{\text{mg}}{\text{L}}\right)\left(0.050 \, \dfrac{\text{L}}{\text{h}}\right)\left(2.0 \, \dfrac{\text{h}}{\text{event}}\right)}{(26 \text{ kg})(25\,550 \text{ d})}$$

$$\times \left(140 \, \dfrac{\text{events}}{\text{yr}}\right)(8 \text{ yr})$$

$$= 0.000\,135 \text{ mg/kg·d}$$

The intake through dermal absorption while swimming is

$$I_D = \frac{CA_s R_D t_E f_E D_t}{(\text{BW})(\text{AT})}$$

$$= \frac{\left(0.8 \, \dfrac{\text{mg}}{\text{L}}\right)(0.94 \text{ m}^2)\left(8.4 \times 10^{-4} \, \dfrac{\text{cm}}{\text{h}}\right)}{(26 \text{ kg})(25\,550 \text{ d})\left(100 \, \dfrac{\text{cm}}{\text{m}}\right)\left(\dfrac{1 \text{ m}^3}{1000 \text{ L}}\right)}$$

$$\times \left(2.0 \, \dfrac{\text{h}}{\text{event}}\right)\left(140 \, \dfrac{\text{events}}{\text{yr}}\right)(8 \text{ yr})$$

$$= 0.000\,021 \text{ mg/kg·d}$$

The total intake is

$$I_I + I_D = 0.000\,135 \, \dfrac{\text{mg}}{\text{kg·d}} + 0.000\,021 \, \dfrac{\text{mg}}{\text{kg·d}}$$

$$= 1.56 \times 10^{-4} \text{ mg/kg·d}$$

The answer is (B).

281. The MCL for trihalomethanes is 0.040 mg/L.

The EPA recommended values for estimating intake are

drinking water ingestion rate, $R_W = 1$ L/d
exposure frequency, $f_E = 365$ d/yr
body weight for a child aged 5 yr to 12 yr, BW = 26 kg

The intake through ingestion with drinking water is

$$I_W = \frac{CR_W f_E D_t}{(\text{BW})(\text{AT})}$$

$$= \frac{\left(0.040 \, \dfrac{\text{mg}}{\text{L}}\right)\left(1 \, \dfrac{\text{L}}{\text{d}}\right)\left(365 \, \dfrac{\text{d}}{\text{yr}}\right)(8 \text{ yr})}{(26 \text{ kg})(25\,550 \text{ d})}$$

$$= 1.8 \times 10^{-4} \text{ mg/kg·d}$$

The answer is (A).

282. The MSDS does not include the MCL, RMCL, or MCLG for the material. These parameters are associated with acceptable concentrations in drinking water and have no direct relationship with workplace safety.

The answer is (B).

283. The MSDS includes information describing chemical exposure pathways into the body, acute and chronic health effects, and medical and first aid treatments for accidental exposure.

The answer is (D).

284. The MSDS would address spill response measures and disposal of wastes generated from spill response, as well as proper storage procedures for the material. Decontamination of wastes generated from spill response do not directly affect workplace safety and would not be included on the MSDS.

The answer is (C).

285. Radon gas is a naturally occurring gas that can cause lung cancer.

The answer is (A).

286. The EPA has regulatory authority for radon gas exposure under the Indoor Radon Abatement Act of 1988.

The answer is (C).

287. The standard unit of measure for radon gas is picocuries per liter of air, pCi/L.

The answer is (B).

288. The level of radon existing in the home and the amount of time spent there are two primary factors influencing risk of developing health problems, including cancer, from radon exposure. This risk is significantly increased for smokers and former smokers.

The answer is (D).

289. The average velocity of the groundwater is

$$v_{aw} = \frac{Ki}{n_e} = \frac{\left(19\ \frac{m}{d}\right)(0.0017)}{0.24}$$
$$= 0.135\ \text{m/d} \quad (0.14\ \text{m/d})$$

The answer is (B).

290. The intrinsic permeability of the soil is

$$k = \frac{K\mu}{\rho g}$$

μ and ρ are the absolute viscosity and density, respectively, of water at 8°C.

$$k = \frac{\left(19\ \frac{m}{d}\right)\left(1.39 \times 10^{-3}\ \frac{kg}{m \cdot s}\right)}{\left(9.81\ \frac{m}{s^2}\right)\left(999.8\ \frac{kg}{m^3}\right)\left(86\,400\ \frac{s}{d}\right)}$$
$$= 3.1 \times 10^{-11}\ \text{m}^2$$

The answer is (B).

291. The hydraulic conductivity with respect to the fuel is

$$K_f = \frac{kg}{v}$$

Assume kinematic viscosity given is for 8°C.

$$K_f = \frac{(1.0 \times 10^{-10}\ \text{m}^2)\left(9.81\ \frac{m}{s^2}\right)\left(86\,400\ \frac{s}{d}\right)}{\left(8.32\ \frac{mm^2}{s}\right)\left(\frac{1\ m}{1000\ mm}\right)^2}$$
$$= 10.2\ \text{m/d} \quad (10\ \text{m/d})$$

The answer is (D).

292. The average velocity of the light non-aqueous phase fuel is

$$v_{af} = \frac{K_f i}{n_e} = \frac{\left(10\ \frac{m}{d}\right)(0.0017)}{0.24}$$
$$= 0.071\ \text{m/d}$$

The answer is (C).

293. For a first-order reaction,

$$\ln \frac{C}{C_o} = -kt$$

$$k = \frac{-\ln \frac{C}{C_o}}{t}$$

At 2 h,

$$k = \frac{-\ln \left(\frac{4.7\ \frac{mg}{L}}{5.1\ \frac{mg}{L}}\right)}{2\ \text{h}} = 0.041\ \text{h}^{-1}$$

At 8 h,

$$k = \frac{-\ln \left(\frac{3.9\ \frac{mg}{L}}{5.1\ \frac{mg}{L}}\right)}{8\ \text{h}} = 0.034\ \text{h}^{-1}$$

At 16 h,

$$k = \frac{-\ln \left(\frac{2.5\ \frac{mg}{L}}{5.1\ \frac{mg}{L}}\right)}{16\ \text{h}} = 0.045\ \text{h}^{-1}$$

At 28 h,

$$k = \frac{-\ln \left(\frac{1.5\ \frac{mg}{L}}{5.1\ \frac{mg}{L}}\right)}{28\ \text{h}} = 0.044\ \text{h}^{-1}$$

$$k_{ave} = \frac{\frac{0.041}{h} + \frac{0.034}{h} + \frac{0.045}{h} + \frac{0.044}{h}}{4}$$
$$= 0.041\ \text{h}^{-1}$$

The answer is (B).

294. No. 1,2,4,5-TCB is a chlorinated aromatic that is not readily biodegradable. At a concentration of 5.1 μg/L, even if it were biodegradable, there would be insufficient substrate to sustain biological activity.

The answer is (D).

295. Gases associated with malodors are typically produced under anaerobic conditions.

The answer is (B).

296. Hydrogen sulfide gas (H_2S) produced by the biological reduction of sulfates is the source of one of the most readily recognizable malodors associated with wastewater treatment and is detectable at very low concentrations.

The answer is (A).

297. From the illustration, the minimum dose required to obtain a free chlorine residual is about 7.7 mg/L.

The answer is (C).

298. From the illustration, a free chlorine residual of 2.0 mg/L requires a dose of about 9.7 mg/L.

The answer is (D).

299. From published atmospheric stability conditions tables, the stability category is F for a wind speed of 2.1 m/s at night with a clear sky.

Using published vertical dispersion coefficient plots, for a 1.4 km distance the standard deviation along the z-axis is $\sigma_z = 17$ m.

Using published horizontal dispersion coefficient plots, for a 1.4 km distance the standard deviation along the y-axis is $\sigma_y = 45$ m.

Assume the standard deviation along the x-axis is the same as the standard deviation along the y-axis, $\sigma_x = 45$ m.

The emission mass is $Q_m = 10$ kg.

The maximum concentration at the residence is

$$C_{1400,0} = \frac{Q_m}{\sqrt{2}\pi^{3/2}\sigma_z \sigma_y \sigma_x}$$

$$= \frac{(10 \text{ kg})\left(10^6 \, \frac{\text{mg}}{\text{kg}}\right)}{\sqrt{2}\pi^{3/2}(17 \text{ m})(45 \text{ m})(45 \text{ m})}$$

$$= 36.9 \text{ mg/m}^3 \quad (37 \text{ mg/m}^3)$$

The answer is (B).

300. The plume moves with the surrounding wind at 2.1 m/s. The center of the plume will arrive at the house in approximately

$$\frac{(1.4 \text{ km})\left(1000 \, \frac{\text{m}}{\text{km}}\right)}{\left(2.1 \, \frac{\text{m}}{\text{s}}\right)\left(60 \, \frac{\text{s}}{\text{min}}\right)} = 11 \text{ min}$$

The leading edge of the plume will arrive before 11 min, and the residents will have to evacuate before then to avoid exposure. Therefore, the time for evacuation is less than 11 min.

The answer is (B).

Resources

ASCE Manual of Practice, No. 62.

Dean, John A. *Lange's Handbook of Chemistry.* New York: McGraw Hill, 1998.

Fetter, C.W. *Contaminant Hydrogeology.* Englewood Cliffs, N.J.: Prentice Hall, 1999.

Freeze, R. Allan, and John A. Cherry. *Groundwater,* 2nd ed. Englewood Cliffs, N.J.: Prentice Hall, 1999.

Kiely, Gerard. *Environmental Engineering.* New York: McGraw Hill, 1996.

Kostecki, P.T. and E.J. Calabrese. *Petroleum Contaminated Soils—Vol 3.* Chelsea, Mich.: Lewis, 1990.

Lide, David R. *Handbook of Chemistry and Physics.* Boca Raton, La.: CRC Press, 2002.

Masters, Gilbert M. *Introduction to Environmental Engineering and Science.* Englewood Cliffs, N.J.: Prentice Hall, 1997.

Metcalf & Eddy, Inc. *Wastewater Engineering.* New York: McGraw-Hill, 2003.

Middlebrooks, E. Joe, et al. *Wastewater Stabilization Lagoon Design Performance and Upgrading.* New York: Macmillan, 1982.

Noll, Kenneth E. *Fundamentals of Air Quality Systems.* Annapolis, Md.: American Academy of Environmental Engineers, 1999.

OSHA Standard 19.10.1000. Air Contaminants.

Peavy, Howard S., Donald R. Rowe, and George Tchobanoglous. *Environmental Engineering.* New York: McGraw-Hill, 1985.

Sawyer, Clair N., et. al. *Chemistry for Environmental Engineering.* New York: McGraw Hill, 2003.

Sincero, Arcadio P. and Gregoria A. Sincero. *Environmental Engineering—A Design Approach.* Upper Saddle River, N.J.: Prentice Hall, 1996.

Steel, E.W., and Terence J. McGhee. *Water Supply and Sewerage.* New York: McGraw-Hill, 1991.

USDA-SCS, Engineering Division. *Urban Hydrology for Small Watersheds,* TR-55. 1986.

USEPA Documents EPA-402-F94-009, EPA-402-R-93-078, EPA-402-K92-001.

USEPA. *Integrated Risk Information System.* Office of Health and Environmental Asessment. www.epa.gov/iris.

Viessman, W., Jr., and G.L. Lewis. *Introduction to Hydrology.* New York: HarperCollins, 1996.

WEF. Existing Sewer Evaluation and Rehabilitation MOP FD-6. 1994.

Visit www.ppi2pass.com today to order these essential books to prepare for the Environmental PE Exam!

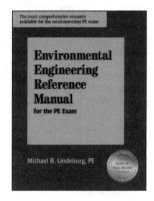

Environmental Engineering Reference Manual for the PE Exam
Michael R. Lindeburg, PE

The **Environmental Engineering Reference Manual** is the most thorough reference and study guide available to engineers preparing for the environmental PE exam. It provides a concentrated review of exam topics, accompanied by hundreds of solved example problems. The text is enhanced by illustrations, tables, charts, and formulas, along with a comprehensive glossary and a detailed index.

Practice Problems for the Environmental Engineering PE Exam: A Companion to the Environmental Engineering Reference Manual
Michael R. Lindeburg, PE

More than 370 practice problems are presented here, corresponding to chapters in the **Environmental Engineering Reference Manual**, so you can work problems as you study. Complete solutions are provided, so you get immediate feedback on your progress and learn the most efficient way to solve problems.

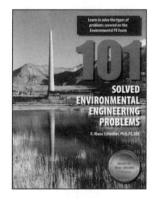

101 Solved Environmental Engineering Problems
R. Wane Schneiter, PhD, PE, DEE

The more problems you solve in practice, the less likely you'll be to find something unexpected on the PE exam. The original problems in this collection are presented in the same multiple-choice format that you'll encounter on the test, and they simulate the difficulty of the PE exam. Every problem includes a complete solution.

For everything you need to pass the exams, go to
www.ppi2pass.com
where you'll find the latest exam news, test-taker advice, the Exam Forum, and secure online ordering.

Professional Publications, Inc.
1250 Fifth Avenue • Belmont, CA 94002
(800) 426-1178 • Fax (650) 592-4519

Source Code BOC 👉 **Quick — *I need additional PPI study materials!***

Please send me the PPI exam review products checked below. I understand any book may be returned for a full refund within 30 days. I have provided my credit card number and authorize you to charge your current prices, plus shipping, to my card.

For the FE Exam
☐ FE Review Manual
☐ Civil Discipline-Specific Review for the FE/EIT Exam
☐ Mechanical Discipline-Specific Review for the FE/EIT Exam
☐ Electrical Discipline-Specific Review for the FE/EIT Exam
☐ Engineer-In-Training Reference Manual
  ☐ Solutions Manual, SI Units

For the PE, SE, and PLS Exam
☐ Civil Engrg Reference Manual   ☐ Practice Problems
☐ Mechanical Engrg Reference Manual   ☐ Practice Problems
☐ Electrical Engrg Reference Manual   ☐ Practice Problems
☐ Environmental Engrg Reference Manual
  ☐ Practice Problems
☐ Chemical Engrg Reference Manual   ☐ Practice Problems
☐ Structural Engrg Reference Manual
☐ PLS Sample Exam

Mail this form to:
PPI, 1250 Fifth Ave., Belmont, CA 94002

These are just a few of the products we offer for the FE, PE, SE, and LS exams. For a full list, plus details on money-saving packages, call Customer Care at 800-426-1178 or visit www.ppi2pass.com.

For fastest service,
Call **800-426-1178**   Fax **650-592-4519**
Web **www.ppi2pass.com**

Please allow two weeks for UPS Ground shipping.

NAME/COMPANY _____
STREET _____ SUITE/APT ____
CITY _____ STATE ____ ZIP ____
DAYTIME PH # _____ EMAIL _____
VISA/MC/DSCVR # _____ EXP. DATE ____
CARDHOLDER'S NAME _____
SIGNATURE _____

Email Updates Keep You on Top of Your Exam

To be fully prepared for your exam, you need the current information. Register for PPI's Email Updates to receive convenient updates relevant to the specific exam you are taking. These will include notices of exam changes, useful exam tips, errata postings, and new product announcements. There is no charge for this service, and you can cancel at any time.

Register at **www.ppi2pass.com/cgi-bin/signup.cgi**

Free Catalog of Tried-and-True Exam Products

Get a free PPI catalog with a comprehensive selection of the best FE, PE, SE, FLS, and PLS exam-review products available, user tested by more than 700,000 engineers and surveyors. Included are books, software, videos, calculator products, and all the NCEES sample-question books—plus money-saving packages.

Request a Catalog at **www.ppi2pass.com/catalogrequest.html**

How to Report Errors

Find an error? We're grateful to every reader who takes the time to help us improve the quality of our books by pointing out an error. It's easy to do. Just go to the Errata Report Form on the PPI website, at **www.ppi2pass.com/Errata/errataform.html**, and tell us about the error you think you've found.

What's more, you can get credit for your contribution to our books. If you're the first person to discover an error, we'll list your name on our website along with the correction. Each errata report is forwarded to the appropriate author or subject matter expert for verification, and valid corrections are added to the Errata section of our website.

But before reporting an error, please check the errata listings on our website. The error you noticed may already have been identified. From our home page at **www.ppi2pass.com**, click on "Errata" (or go directly to **www.ppi2pass.com/Errata/Errata.html**). It's a good idea to check this page before you start studying and periodically thereafter.

Look for your errata on our website on or around the first of each month. That's when valid errata are posted. During the exam months of April and October, new errata are posted as they are verified, up until the day of the exam.

You may also fax errata to us at (650) 592-4519 or mail them to Professional Publications, Inc., c/o Editorial Errata Department, 1250 Fifth Avenue, Belmont, CA 94002, USA. Be sure to include your name, the book title, the edition and printing numbers, the page number(s), and any other information that will help us locate the error(s).

PROFESSIONAL PUBLICATIONS, INC.
800-426-1178 • Fax 650-592-4519
errata@ppi2pass.com